HISTOLOGY AND CYTOLOGY

An AVI Series
AVI Publishing Company

LABORATORY MANUAL OF HISTOLOGY AND CYTOLOGY

Stanley Lawrence Lamberg, B.S., M.A., M.S., Ph.D.

Registered M.T.
Professor
Department of Medical Laboratory Technology
State University of New York
Agricultural and Technical College
Farmingdale, New York

Robert Rothstein, B.A., M.A.

Registered M.T.
Chairman
Division of Human Services
Professor
Department of Medical Laboratory Technology
State University of New York
Agricultural and Technical College
Farmingdale, New York

AVI PUBLISHING COMPANY, INC.
Westport, Connecticut

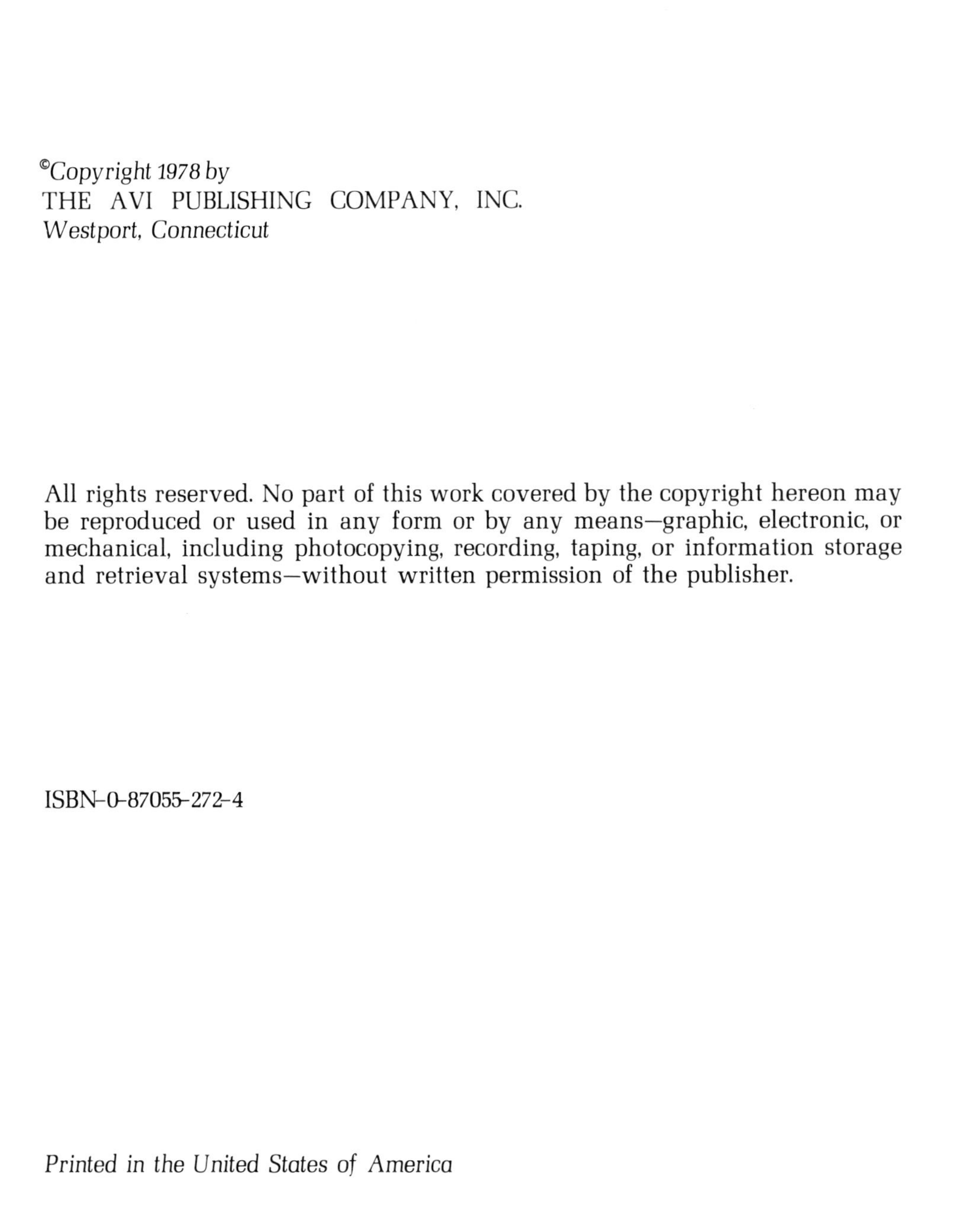

ISBN-0-87055-272-4

Printed in the United States of America

Preface

Histology is a course normally offered in the upper two years of a baccalaureate degree in biology or in professional degree programs.

The authors of this manual have been offering this course to students enrolled in a two year degree curriculum in Medical Laboratory Technology.

It was, therefore, the purpose of the authors in writing this manual to incorporate techniques and skills basic to any level of educational sophistication and to include relevant theoretical considerations that would lead to comprehension of the techniques.

The exercises were designed to be progressive and cumulative in nature and it is assumed that the manual would be used in addition to the traditional lecture component accompanied by any one of the suitable available textbooks in accord with the instructor's desires and emphasis.

Automated tissue processing and staining procedures as well as traditional manual techniques are included to cover the broad spectrum of teaching situations where the amount of equipment is highly variable. Cryostat techniques and Papanicolaou staining methods are included. The authors lead the student step by step from general principles of preparation of tissues to highly specific staining techniques.

It is the hope of the authors that after completion of the exercises included, a student would be capable of the total preparation of excellent slides with a comprehensive understanding of every step in the procedure.

It is also assumed by the authors that tissue recognition and identification, which are usually included in the histology/cytology courses, be best left to the discretion of the instructor as to time, mode and scope of this component of the course.

The questions at the end of each exercise should serve as an ongoing review process for the student.

STANLEY LAWRENCE LAMBERG
ROBERT ROTHSTEIN

April 1978

Acknowledgments

The authors gratefully acknowledge the assistance of Mr. John Frost for his artwork, and Marsha Mason for preparation of the manuscript.

The authors would also like to thank Dr. Donald K. Tressler, AVI Publishing Company, for his encouragement, and Barbara Flouton and Arlene Hoeppner of the AVI editorial department for their assistance.

Finally, the authors thank their students who expressed the need for this manual and Dr. George G. Cook and Dr. Norman Desrosier for bringing the authors and the AVI Publishing Company together to accomplish not only the publication of this manual but also the other four manuals in the series *Functional Medical Laboratory Technology: A Comprehensive Series of Manuals.*

STANLEY LAWRENCE LAMBERG
ROBERT ROTHSTEIN

Contents

PREFACE

ACKNOWLEDGMENTS

1 General Principles for the Preparation of Tissues for Histological Study

Selecting and Obtaining Tissue; Fixation; Washing; Dehydration; Clearing; Infiltration; Embedding; Microtomy; Attaching Sections to Slides; Staining; Mounting a Cover Slip 1

2 Basic Procedures for Preparation of Tissue by the Paraffin Method

Selecting and Obtaining Mammalian Tissue; Fixation (Aims and Objectives of Fixation, Nature of Fixatives, Routine and Special Fixatives, Choice of Fixatives); Washing Tissues 8

3 Dehydration, Clearing, Infiltration, Embedding and Routine Timing Schedule for Manual Technique 24

4 Automatic Tissue Processing

Components and Principles of the Ultra I and Ultra II; Reagents Used for Automatic Tissue Processing 35

5 Microtomy or Sectioning

Principles of Operation of the Rotary Microtome; Use and Care of Microtome Knives; Razor Blade Technique; Preparation of Paraffin Tissue Block; Adjustment of the Rotary Microtome and Sectioning; Problems Encountered and Remedies Utilized in Paraffin Sectioning 43

6 Spreading the Sections and Attachment or Mounting of Sections to Glass Slides 58

7 General Staining Procedure for Paraffin Infiltrated and Embedded Tissue

Nuclear Stains; Cytoplasmic Stains; Equipment and Procedure for Manual Staining; Automated Staining Technique 63

8 Mounting of the Cover Slip and Types of Mounting Media; Labeling and Cataloguing the Slides 82

9 Theoretical Aspects of Staining

Types of Stains; Chemical Staining Action; Progressive and Regressive Staining 89

10 Specific Staining Procedures

Connective Tissue Fibers, Elastic Tissue, Carbohydrates, Special Cells, Nucleic Acids; Exfoliative Cytology (Papanicolaou Technique) 99

11 Cryostat Sectioning, Technique and Staining 130

REFERENCES

1

General Principles for the Preparation of Tissues for Histological Study

The primary objective of histology is to study the structural relationships of the tissues of the organism to gain an insight into how the living organism functions. Ideally, we should study living cells and tissues and there are several methods available to accomplish this. Unfortunately these methods are quite sophisticated and complex and not sufficiently simplified for routine histologic and pathologic studies. Furthermore, examination of cells and tissues in the living state is limited by the transparency of the cellular and tissue components, as the components are not sufficiently differentiated and contrasted from each other. In addition, the thickness of the cells and tissues in the living state interferes with passage of light. The living cells and tissues die quickly when excised from the body and postmortem changes rapidly set in rendering them unfit for study. For these reasons and others, it is not feasible to use living cells and tissues to study the organization of cellular and tissue components and to relate this organization to function. Therefore, for routine histologic and pathologic study, tissues which have been killed chemically, preserved and fixed, cut into thin sections, and stained are used. Chemical preservation eliminates postmortem changes for the most part. Thin sections overcome the interference with passage of light. Staining increases contrast of the tissue components so that they can be resolved and recognized in the light microscope.

The ideal histological technique should "result in minimum deviation"[1] of the tissue from the living state, but still "permit maximum resolution of the components."[1] This condition is never obtained, as preparation of tissue by any histological technique usually leads to altered arrangement of the tissue components from the living state. The artificial appearance of the tissue due to the procedure of preparation is known as artifact.

Several terms have been used so far that will now be defined:

1. **Histology.**—The microscopic study of structure of normal tissue.

2. **Pathology.**—The study of diseased tissues and conditions and processes of a disease.

3. **Tissue.**—The term tissue, when used in context of a histological technique, is used in a nonspecific way. A piece of tissue is anything surgically removed from the body of an organism and processed through the histological procedure.

[1]Bloom, W., and Fawcett, D.W. 1968. *A Textbook of Histology, 9th Edition.* W.B. Saunders Co., Philadelphia.

4. **Section.**—A section is a thin slice of tissue laid flat on a glass slide.

5. **Postmortem Changes.**—Postmortem changes are degenerative changes of the cells and tissues of the body rendering them subject to significant alteration when used for microscopic study.

Routine Production of Histological Slides

The technique for routine production of histological slides is the paraffin histological or microtechnique. This procedure involves 15 basic steps in preparation of a histological slide. The procedure can be carried out manually, although several steps of the procedure are now automated.

The following is a brief description of the procedure to give the student an overview of all the steps involved in the production of a finished histological slide.

1. **Selecting and Obtaining Tissue.**—
   a. Sacrifice an animal and dissect out small pieces of tissue as soon as possible after the death of the animal. The anesthetic commonly used is either ether or chloroform. After the animal is sacrificed, all the desired parts are excised and placed in a fixative.

   b. Biopsy and autopsy specimens of animal (human) tissue. The physician or veterinarian removes the tissue specimens; they are promptly placed in a fixative and brought to the histology/pathology laboratory for subsequent processing.
      (1) *Biopsy specimens* are excised small pieces of living tissue obtained from a surgical procedure. They are immediately placed in a fixative and are usually free of postmortem changes.
      (2) *Autopsy specimens* are excised pieces of tissue removed from the body after death. The tissue will exhibit varying degrees of postmortem changes depending on length of time before fixation occurred.

      Once a tissue is removed from an organism, it is given a code number which is carried along with the tissue during all the steps of the procedure.

2. **Fixation.**—To prevent degeneration and preserve the tissue as close as possible to the living state, the tissue, after removal from the body, is placed in solutions called fixatives for specified periods of time. Fixatives are usually a combination of two or more chemical reagents mixed together, except for formalin. Fixatives are generally named after the person who first formulated the mixture of reagents, for example, Bouin's or Zenker's fixative. Different fixatives are used to preserve specific components of the tissues. Thus before tissue is removed from the animal, thought should be given to the correct choice of fixative, since use of the wrong fixative will destroy or distort the tissue component that you want to demonstrate.

   Some common chemical reagents used in combination to form fixatives are:

   Acetic acid
   Acetone
   Ethyl alcohol
   Formaldehyde (formalin)
   Mercuric chloride
   Picric acid
   Potassium dichromate
   Trichloroacetic acid

The fixatives do two things to preserve the tissues.

a. **Immediately Kill the Tissues.**—As the fixative penetrates into the tissue, the cells of the tissue are quickly killed. This preserves the structural integrity of the cells and tissues by interrupting the living processes and prevents post-mortem changes, such as autolysis. As a result, cellular and tissue structures are stabilized as similarly as possible to the condition they were in life with a minimum of change.

b. **Harden the Tissues.**—As the fixative comes into contact with the structural components of cells and tissues, especially the proteins, they become insoluble and precipitate out in place. This hardens the tissue. The hardening of the tissue protects against shrinkage and distortion during subsequent histological steps. The reason then for dissecting out small pieces of tissue is to enable the fixative to penetrate quickly and thoroughly to accomplish the above fixation objectives.

3. **Washing.**—After the tissue is fixed for the proper length of time, excess fixative is washed out to prevent over-fixation. Washing also removes substances in the fixative which might interfere with the subsequent processing. Since most fixatives are aqueous solutions, the washing is usually carried out for a specific period of time in cold running tap water.

4. **Dehydration.**—Tissue contains much "free" water. As paraffin, which is to be used later in the procedure (Steps 6 and 7), does not mix with water, the water of the tissue is removed by dehydration. In this step, tissue is placed into a series of gradually increasing concentrations of alcohol, usually ethyl alcohol (30, 50, 70, 80, 95, and 100%), for specific periods of time.

5. **Clearing (Removal of Alcohol).**—Paraffin is not soluble in alcohol. The alcohol is replaced by a substance in which paraffin is soluble, such as xylene, toluene, or benzene. These substances are called clearing agents. The tissue is placed in several changes of clearing agent for specific periods of time. Clearing also increases hardening of the tissue and makes the tissue translucent.

6. **Infiltration.**—In order for tissue to be sectioned into thin slices, it must be supported by permeating it with a medium that will give it a proper consistency for sectioning. Paraffin is commonly used as an infiltration medium. After clearing, the tissue is placed in several changes of melted paraffin for a specific period of time. The infiltration occurs under vacuum and heat. The vacuum helps infiltrate the paraffin through the tissue and the heat keeps the paraffin liquified. The paraffin permeates, impregnates or infiltrates the tissue, replacing the clearing agent.

7. **Embedding.**—The paraffin infiltrated tissue is placed into a mold into which liquid paraffin is poured. The paraffin is now allowed to harden around the tissue into a solid mass or "block" conforming to the shape of the mold. The solid paraffin "block" containing the tissue is then removed from the mold. The tissue, having gone through the infiltration and embedding steps, is now of proper consistency for sectioning.

8. **Microtomy or Sectioning.**—The solid paraffin "block" containing the tissue is sliced into very thin sections on an instrument called a rotary microtome. The paraffin tissue "block" is attached to a tissue holder, which is then placed on the microtome. The range of thickness of the sections in paraffin technique is 3–10 microns (μ) with an average section thickness of 6 μ. In some cases 5 μ sections are desirable. The paraffin sections as they are cut adhere to one another to form a ribbon. Movement of the rotary handle of the microtome moves the paraffin "block" closer to the microtome knife; as a result the sections are cut from the face of the "block" and come off as a ribbon.

9. **Attaching or Mounting Sections on Slides and "Spreading" the Sections.**—

   a. **Warming Plate Method.**—A thin coating of some type of an adhesive, such as albumin or glycerin-albumin is rubbed on one surface of a glass slide. A drop or two of distilled water is placed on the slide and several paraffin sections of the ribbon containing the tissue are floated on the water. The slide is placed on a slide warming plate (45°–50°C) so as to soften the paraffin and flatten out the folds that formed in the tissue on sectioning. When the tissue sections have flattened, the slide is removed from the warming plate and air dried to firmly adhere tissue sections to the glass slides.

   b. **Flotation Method.**—An alternate method of attachment of sections to slides is to float the paraffin ribbon containing the tissue sections onto the surface of a warm water bath (45°–50°C) containing gelatin. The gelatin acts as an adhesive for attachment of sections to the slides. The heat of the water bath softens the paraffin to remove folds in the tissue sections. The ribbon is divided into small units of 3–5 sections each. The slide is placed under the sections and the sections are picked up and centered on the slide. The excess water is drained off and slides with attached sections are placed on a slide warming plate (45°–50°C) until sections adhere to slide. The tissue sections are then air dried to firmly attach them to the glass slide.

10. **Preparation of Paraffin Sections for Staining (Deparaffinization and Hydration).**—Since many stains utilize water as the solvent, the paraffin in which the tissue is embedded must be removed. This is accomplished by placing the slide to which the tissue sections are attached into a clearing agent to dissolve the paraffin, then into absolute alcohol to remove the clearing agent. From the absolute alcohol the tissue sections are placed successively into weaker concentrations of alcohol and finally water. The tissue sections are now ready to be stained.

11. **Staining.**—The staining methods developed increase the contrast among the tissue components by use of two or three stains which selectively color the different tissue components. The colors obtained are not necessarily indicators of the chemical nature of the tissue component being stained. The application of several stains to bring out differences among the tissue components is referred to as staining and counterstaining.

12. **Theoretical Aspects of Staining.**—These will be discussed in great detail in the subsequent exercise on staining.

13. **Dehydration and Clearing of Stained Tissue Sections.**—After tissue sections are stained, they are converted into permanent preparations. The tissue sections are rinsed in water to remove excess stain and sometimes to enhance staining properties. The water is removed from the tissue sections by passing them through an increasing series of alcohols to absolute alcohol and then into a clearing agent such as xylene. The clearing agent makes the stained tissue sections transparent. The clearing agent is also the solvent for the mounting medium for adhering the cover slip.

14. **Mounting.**—A drop of mounting medium, such as Permount[2], is put on the section and a cover slip is mounted over the section. As the solvent of the mounting medium evaporates, the mounting medium hardens and encloses the tissue sections in an almost solid substance and attaches the cover slip to the slide. After the cover slip is firmly attached, the slide can be handled without damage to the tissue sections. A suitable mounting medium dries quickly, has the same refractive index as glass (allows light to pass at the same rate as glass) and will not fade the staining of the tissue sections with time.

15. **Cleaning, Labeling the Slide and Cataloguing the Slide in the Slide Box Register.**—These are the final steps in the preparation of a histological slide.

    a. **Cleaning.**—After the mounting medium has hardened, excess mounting medium from slide and cover slip is removed by scraping with a razor blade and using a cotton swab moistened with xylene.

    b. **Labeling.**—The pertinent identification information is recorded on the slide label according to the procedure used in the particular laboratory.

    c. **Cataloguing the Slide in the Slide Box Register.**—The slide is registered in the catalogue so it may be recovered for further study at a later time.

Automation

Those steps of the procedure which are automated are Steps 2, 3, 4, 5, 6, 7 (but not completely), 10, 11, 13, and 14 (within the near future). There are several automated instruments that can carry out these steps in the procedure such as the Autotechnicon[3] and Ultra-Autotechnicon[3].

Action of Fixatives and Stains

In order to demonstrate the various cellular and tissue components, the proper choice of fixative must be made, otherwise structures will be destroyed. Furthermore, in order to observe these structures, the correct stain must then be supplied.

For example, if you want to demonstrate mitochondria in tissue cells, the tissue must be fixed in a neutral fixative, such as neutral buffered formalin to preserve the mitochondria, and then stained with the proper stain, such as Heidenhain's iron hematoxylin, so the mitochondria can be demonstrated. If an acidic fixative, which destroys mito-

[2]Fisher Scientific Co., Inc., Pittsburgh, Pa.
[3]Technicon Instrument Corp., Tarrytown, N.Y.

chondria, were to be used, the mitochondria would not be observable upon staining with iron hematoxylin.

On the other hand, the mitochondria would not be observable if the incorrect stain were to be used, even though these structures were preserved by proper fixation. If hematoxylin and eosin were used after proper fixation, the hematoxylin, a blue stain, would stain the nucleus blue; eosin, a red-pink stain, would stain all cytoplasmic structures and intercellular substances red-pink. Nothing of the cytoplasmic components of the cell could be seen, as the eosin does not differentially stain the cytoplasmic components. Thus, the mitochondria are not visualized by the stain, although they are preserved by fixation. It is essential then to employ the proper fixative and stains if a specific tissue component is to be demonstrated.

These significant features of tissue preparation will be discussed in detail in successive exercises of this manual.

QUESTIONS

Cover answers with a piece of paper. Answers appear at end of questions.

(1) It is not feasible to use living cells to study the structural relationships of the tissue of the organism primarily because
(a) It can't be done
(b) Tissues undergo rapid postmortem changes when excised from the body
(c) They are not truly representative because of distortion
(d) They cannot be studied with the light microscope

(2) "The study of diseased tissues and conditions and processes of a disease" is a definition of
(a) Cytology
(b) Histology
(c) Pathology
(d) Artifaction

(3) Small excised pieces of living tissue obtained from surgical procedures are called
(a) Biopsy specimens
(b) Autopsy specimens
(c) Cytology specimens
(d) Postmortem specimens

(4) The following: acetic acid, picric acid, formalin, potassium dichromate, ethyl alcohol, are referred to as
(a) Dehydrants
(b) Infiltrations
(c) Stains
(d) Fixatives

(5) Dehydration is usually accomplished by placing tissue in
(a) Xylene
(b) Paraffin
(c) Toluene or benzene
(d) Various concentrations of ethyl alcohol

(6) The step following dehydration is referred to as
(a) Clearing
(b) Embedding
(c) Infiltration
(d) Sectioning

(7) Slicing a paraffin block containing a piece of tissue is called
- (a) Clearing
- (b) Embedding
- (c) Microtomy
- (d) Flotation

(8) "Mounting" refers to
- (a) Placing cover slip over stained tissue
- (b) "Capturing" tissue on slide
- (c) Removing paraffin from tissue
- (d) Another term for staining

(9) Which of the following is <u>not</u> a characteristic of a suitable mounting medium?
- (a) Same refractive index as glass
- (b) Will not fade
- (c) Dries quickly
- (d) Is slightly colored

(10) The range of thickness with paraffin technique is
- (a) 1–2 microns
- (b) 20–25 microns
- (c) 3–10 microns
- (d) 15–20 microns

Answers

(1) b
(2) c
(3) a
(4) d
(5) d
(6) a
(7) c
(8) a
(9) d
(10) c

2

Basic Procedures for Preparation of Tissue by the Paraffin Method

I. SELECTING AND OBTAINING MAMMALIAN TISSUE

A. Sacrificing the Animal

Before sacrificing the animal, all equipment should be at hand. This includes sacrificing jar (a large desiccator jar with a tight lid with vacuum grease spread on edge of lid to exclude air is sufficient); anesthetic (ether or chloroform); dissecting instruments and pan; 0.85% physiological saline (NaCl) solution; paper toweling; beakers or vials labeled and filled with proper fixative; and tissue capsules or cassettes to hold the tissue specimens, coded with a code number (see Step D).

The animal should be sacrificed as quickly and humanely as possible. Place the animal into the sacrificing jar into which has been placed a wad of cotton soaked with the anesthetic. Place animal on a wire mesh above the wad of cotton; secure the lid. After several minutes, the animal should have expired. After the animal is dead, remove it from sacrificing jar. Place the animal on its back on a towel-covered dissecting pan, immediately open the animal, and flood the animal's interior with the 0.85% NaCl. The purpose of the 0.85% NaCl is to prevent drying out of the tissues while dissecting tissue specimens.

The tissue specimens should be quickly removed from the animal and put into the fixative. Tissues should be fixed from the living state. If fixation is not rapidly accomplished, the tissue dries and postmortem deterioration sets in. Postmortem changes result from two processes:

1. **Autolysis.**—Enzymes within the lysosomes of the cells are liberated which break down the cellular and tissue components. This is enzyme dissolution from within.

2. **Microbial Putrefaction.**—Breakdown of the cells and tissues by bacterial and mold enzymatic activity. This is enzyme dissolution from without. The enzyme dissolution by either of the two processes leads to postmortem changes.

Fixation reduces the postmortem decomposition to a minimum.

B. Biopsy, Surgical and Autopsy Specimens

Tissues obtained from human sources are either biopsy, surgical or autopsy specimens. These tissue specimens should be processed in the same manner as tissue specimens excised from a sacrificed animal.

C. Preparation of Tissue for Fixation

1. **Size of Tissue.**—Remove small pieces of tissue (0.5 to 2 cm thick) for manual technique. The thickness of tissue for processing by automated systems is between 1 and 3 mm. The small size of the tissue allows for rapid and complete penetration of fixative and more perfect preservation of the tissue components as they were in life. When cutting the tissue use a sharp scalpel. The squeezing action of scissors crushes and tears the tissue. Before the tissue is placed in the fixative, rinse tissue briefly in 0.85% NaCl to remove adhering blood, as blood will retard fixation. Never allow tissue to become dry.

2. **Volume of Fixative to Use.**—Place tissue in a volume of fixative at least 15–20X the volume of the tissue. If tissue is 1–2 cm thick, use 10–20 ml of fixative. Remove tissue rapidly from the animal, place tissue in a coded tissue capsule and immediately into the fixative. Stir fixative gently to make sure fixative reaches all surfaces of the tissue and tissue capsule does not stick to bottom or sides of the container.

D. Assigned Code Number

Assign a code number for each tissue received in numerical order. This code number must remain with the tissue throughout its preparation. If a tissue is improperly coded or code number is applied to the wrong tissue a serious error could result, especially in making a diagnosis for a specific patient.

II. FIXATION

Fixation is the most important step in the preparation of tissue for microscopic examination, since fixation changes the physical and chemical state of the tissue. Fixation determines the morphological status of the tissue and subsequent staining reactions that could be carried out on the tissues. Before choosing a fixative it is necessary to consider for what purpose the tissue is going to be used. To demonstrate general morphological relationships in the tissue, a general or routine fixative could be used. If specific cell types or cell constituents are to be demonstrated, special fixatives are employed in order not to destroy these structures. Furthermore, the choice of fixative must be compatible with the later staining steps. In detailed histological study of tissue a general fixative is used first to gain an overall view of morphology, the special fixatives are used to study specific components of the tissues. Generally, fixatives are chemical solutions, although there are physical means for fixation. No single chemical reagent has all the requirements of an ideal fixative. Usually two or more reagents are combined to take advantage of the specific effect of each to fulfill the fixative objective of preserving the tissue structure in a state as close as possible to the living condition.

A. Aims of Fixation

Aims are penetration of fixative into the tissue to kill and harden the tissue; prevention of postmortem changes and alteration of the tissue during the histological procedure.

B. Objectives of Fixation

To accomplish the aims, the following objectives are involved in the fixation step:

1. **Rapid Penetration of the Fixative to Immediately Kill the Tissue and to Minimize Postmortem Changes in the Tissue.**—Tissues as they are immersed in fixative should be killed and fixed nearly simultaneously. The fixation begins at the periphery and proceeds inward. The rate of penetration will determine how much postmortem change occurs in the center of the tissue before the fixative reaches this region. Penetration rates of fixatives differ greatly. There is no relationship between molecular weight and rate of penetration or the number of reactive atoms on a molecule and rate of penetration. If tissue excised from the animal is too thick, it will be improperly fixed at the center and postmortem changes will be most evident there.

2. **To Preserve the Relations of the Tissue Components as Closely as Possible to What They Were in Life.**—The fixatives kill the tissue by denaturation of the cell's chemical contents, especially the proteins and nucleic acids. Denaturation is a change in the chemical, physical and biological properties of the proteins caused by alteration in structure or composition as the reagents of the fixative combine with the proteins. As a result, the proteins usually coagulate, i.e., become insoluble, with concurrent precipitation. In this manner, the molecules are stabilized and they remain in the original position.

3. **Harden the Tissue.**—During the process of coagulation, the cell contents harden. The hardening allows the tissue to resist any further changes in its architecture during subsequent histological steps. Thus, fixation protects the tissue against shrinkage and distortion as the tissue is processed.

4. **Improve Staining Potential of the Tissue by Bringing Out Differences in Refractive Indexes of the Various Components of the Tissue.**—Some reagents of the fixatives enhance subsequent staining reactions by providing a linkage between a tissue component and the stain. This is known as mordanting.

C. Nature of Fixatives

1. **Chemical Reagents of the Fixatives.**—As no single chemical has qualities of an ideal fixative, several chemical reagents are mixed together to take advantage of their fixing qualities, and, except for formalin, the most efficient fixatives are mixtures of several chemical reagents. The number of chemical reagents used in fixatives is very small. There are several ways of classifying these reagents.

   a. **Coagulant Versus Noncoagulant Fixing Reagents.**—Some of the reagents in fixatives denature the proteins of the tissues and coagulate them. These chemicals are called coagulant chemical reagents. Other chemicals used in fixatives denature the proteins without coagulating them. These are termed noncoagulant chemical reagents. Fixatives generally contain both types.

(1) *Coagulant Fixing Reagents.*—

(a) Advantages.—

1. The coagulant chemical reagents change the fine fibrous network or meshwork of proteins of tissues into a coarse network. This does not destroy the structure at the microscopic level, but permits the paraffin to easily infiltrate the interior of the tissue to form a tissue of proper consistency for sectioning.

2. Another important advantage is that some coagulant chemical reagents strengthen the chemical bonds within and between proteins against breakdown during later histological steps.

(b) Disadvantages.—

1. The coagulant chemical reagents form too coarse a network for best histological and cytological details.

2. They induce the formation of artificial structures in the tissue, known as artifacts.

Some coagulant fixing reagents are mercuric chloride, picric acid, chromium trioxide and ethanol.

(2) *Noncoagulant Fixing Reagents.*—

(a) Advantage.—The noncoagulant chemical reagents produce fewer artifacts, since they do not drastically change the fine network of proteins of tissues. This leads to good histological and cytological details.

(b) Disadvantage.—Since these chemical reagents do not alter the fine network of protein, paraffin does not easily infiltrate into the tissue and tissue has a poor consistency for sectioning.

Some noncoagulant fixing reagents are formaldehyde, potassium dichromate, acetic acid and osmium tetroxide.

b. Additive Versus Nonadditive Fixing Reagents.—

(1) *Additive Fixing Reagents.*—The molecules of the reagent combine with side groups of the protein in a chemical bond. Additive fixation does not necessarily mean that proteins have to coagulate, as some of these chemicals do not coagulate proteins.

(2) *Nonadditive Fixing Reagents.*—The molecules of the reagent do not combine with the proteins in a chemical bond. Nonadditive fixation does not necessarily mean that proteins do not coagulate; some of these chemicals do coagulate the proteins.

2. Properties of Chemical Fixing Reagents

The classification used in the following discussion is based upon coagulant versus noncoagulant fixing reagents.

a. Coagulant Fixing Reagents.—

(1) *Mercuric Chloride.*—This heavy metal salt is used as a 7% aqueous solution or as an alcoholic solution in which it is more soluble (33%). It strongly coagulates the protein as it penetrates at a moderate rate through the tissue. Chemically it is an additive reagent, since it forms chemical bonds or cross linkages between side groups of the protein. This moderately hardens the tissue. Mercuric chloride shrinks the tissue very slightly and distorts the tissue less than other coagulant reagents. It preserves the cytoplasmic components extremely well. Mercuric chloride fixes proteins by coagulation. It does not fix nucleic acids well, so chromosomes are poorly fixed, although nucleoli are distinct. Lipids are basically untouched by this reagent and it is not known to fix carbohydrates, except mucopolysaccharides such as those in mucin.

Tissues fixed in mercuric chloride stain brilliantly as it acts as a mordant for stains.

A disadvantage of mercuric chloride is the appearance of fine black precipitated deposits throughout the tissue. These deposits are removed by a Lugol's iodine-thiosulfate post-fixation treatment.

Another disadvantage is that mercuric chloride fixed tissue cannot be used for frozen or cryostat sectioning as mercuric chloride inhibits freezing of the tissue.

NOTE: Mercuric chloride is poisonous if absorbed through the skin and should be handled with care!

(2) *Picric Acid.*—This yellow crystalline substance is only slightly soluble in water as it forms a saturated aqueous picrate solution of about 1%. In alcohol it is soluble to about 9%. The reagent crystallizes on the bottom of the stock bottle and will go into solution upon addition of more solvent to the stock as the saturated solution is employed in various fixatives.

NOTE: Picric acid must never be kept as a dry powder for it is highly explosive in this state!

Picric acid is a coagulant reagent of proteins, especially nuclear proteins. It is an additive reagent, as the picrate forms cross linkages between the side groups of the protein, but it does not harden the tissue. Picric acid leaves the nucleic acid, DNA, in solution, thus preserves but does not fix the chromosomes. It does not dissolve lipids.

It does not in general fix carbohydrates, but it does fix glycogen by inhibiting solution of the glycogen as the glycogen binds to the protein. Picric acid is an excellent mordant for stains.

Its disadvantages are that it penetrates slowly and causes extreme shrinkage of the tissue. Shrinkage is counteracted by using acetic acid in combination with picric acid. Furthermore, as picric acid is yellow in color, excess picric acid must be removed from the tissue before staining. This is accomplished by treating the tissue with 50–70% alcohol during dehydration to wash out the excess picric acid or by treating the tissue sections in 70% alcohol containing a few drops of lithium carbonate.

(3) *Chromium Trioxide or Chromic Acid.*—This is used in an aqueous solution of 0.5–2%. In water it forms chromic acid. It is a powerful protein coagulant. It chemically acts as an additive reagent. It penetrates slowly, hardens moderately, but shrinks tissue considerably. It coagulates nucleic acid and thus is a good fixative for chromosomes. It oxidizes lipid and makes lipid insoluble in lipid solvents. Thus it is also a good fixative for lipid. It also oxidizes carbohydrate, making it less likely to be dissolved out of the tissue during subsequent histological steps, but it does not fix the carbohydrate. This reagent leaves the cytoplasm coarsely granulated, but it does not fix the cytoplasmic components. Chromic trioxide is a strong oxidizing agent and should generally not be mixed with alcohol or formalin, which reduces it, while they are oxidized. Excess reagent must be washed out of the tissue before dehydration in alcohol, as the chromium trioxide will be reduced to chromic oxide, an insoluble green precipitate which is resistant to removal by acid or other reagents.

(4) *Ethanol (or Ethyl Alcohol).*—It is used undiluted or absolute. It easily mixes with water or xylene. It coagulates many proteins, but acts chemically in a nonadditive manner. Ethanol penetrates moderately, but shrinks, distorts and hardens tissues excessively. It precipitates nucleic acid and glycogen without fixing them; thus they are not stabilized and may even be dissolved out. Some lipids are soluble in ethanol; others are not. Ethanol frees the lipid from combination with proteins. It leaves the cytoplasm coarsely coagulated and destroys cytoplasmic components; and it leaves the chromosomes indistinct and shrinks the nucleoli. Since ethanol can be oxidized, it should not be mixed with oxidizing fixing reagents such as chromium trioxide.

b. Noncoagulant Fixing Reagents.—

(1) *Formalin (Formaldehyde).*—This is normally a gas which is soluble in water to 40%. This 40% solution is considered to be concentrated, or 100% formalin. The 100% formalin is diluted usually with buffered phosphate salt solution (pH 7.5–8.0) in the proportion of 10 parts formalin to 90 parts diluent to form a 10% neutral buffered formalin solution. The 10% formalin is the usual concentration at which this fixing reagent is used.

Formalin is a noncoagulant fixing reagent. Chemically, it reacts with the side groups of the amino acids of the proteins to form chemical bonds between adjacent protein chains. The pH of the formalin solution should be at 7.5–8.0, as the amount of formalin bound to the protein decreases as pH rises above 10. Formalin penetrates the tissue rapidly and causes slight shrinkage. Further shrinkage of the tissue occurs when the tissue is subjected to the subsequent steps in the histological procedure. Formalin appears to have a beneficial hardening effect on the tissue, although many histologists consider it to be a "soft" fixing reagent. Formalin does not fix nucleic acids but dissolves them. Most lipids are preserved, but not fixed, by formalin. Formalin does not fix carbohydrates, but as the formalin fixes the protein, the carbohydrate (glycogen) is not easily dissolved out by subsequent treatment since the protein maintains the carbohydrate within the tissue.

Due to the type of chemical reaction between formalin and protein, formalin enhances nuclear staining compared to cytoplasmic staining.

Unbuffered formalin reacting with the hemoglobin of the red cells will form an artifact type of pigment. Use of neutral buffered formalin will prevent this pigment formation.

(2) *Potassium Dichromate.*—This reagent is used in a 2.5–5% aqueous solution, but is insoluble in alcohol. At pH 3.4–3.8 or above, it is a noncoagulant reagent which penetrates slowly and leaves the tissue soft and fairly swollen. Subsequently, paraffin does not infiltrate sufficiently and tissue has a poor consistency for sectioning. Potassium dichromate dissolves nucleic acids and leaves chromosomes and nucleoli scarcely visible. Its importance in fixation is chiefly for its effect on lipids, as it renders lipids insoluble in lipid solvents and so preserves mitochondria. The chromium trioxide, formed from the dichromate ion, partially oxidizes certain lipids, and the chromium ion also chemically combines with certain other lipids. It does not fix any carbohydrate.

The potassium dichromate can become a coagulant reagent upon being acidified. It then resembles chromium trioxide and penetrates rapidly. The pH of the solution thus affects the mode of fixation by this reagent. Potassium chromate by itself is almost useless during fixation, and should be used in combination with other fixing reagents; it is an effective mordanting reagent during staining. Potassium dichromate must be washed out of the tissue with water before tissue is dehydrated with alcohol, as alcohol reduces the dichromate ion to chromic oxide. The chromic oxide precipitates out as green crystals in the tissue.

(3) *Acetic Acid.*—This is usually used in its concentrated form as glacial acetic acid. The acetic acid is a noncoagulant, nonadditive fixing reagent which penetrates rapidly and does not harden the tissue. It leaves the tissue softer than any other fixing reagent. It prevents the hardening that occurs during dehydration in alcohol. Acetic acid dissociates nuclear proteins from the DNA and fixes both components

by precipitating them; thus the chromosomes are well preserved. However, acetic acid does not fix cytoplasmic proteins, but hydrates them. Acids in general, and acetic acid in particular, cause tissues to swell by breaking down the cross linkages between the protein collagen molecules. This is advantageous as it counters shrinkage of tissue caused by other fixing reagents, such as picric acid and mercuric chloride. This is the main reason for its use as a component in fixatives.

Acetic acid does not fix lipids, and in concentrated form it dissolves them, thus dissolving many of the cytoplasmic components. It neither fixes nor destroys carbohydrates.

Acetic acid should not be used with potassium dichromate as the dichromate will behave as chromium trioxide.

(4) *Osmium Tetroxide.*—This is used as a 1% aqueous solution. The osmium tetroxide is a noncoagulant reagent that chemically acts as an additive reagent, which renders the protein no longer coagulable by alcohol. It penetrates the tissue poorly, does not change the volume of the tissue, and leaves tissue soft but friable, or crumbly, in the paraffin so tissue sections poorly. Osmium tetroxide does not coagulate nucleic acid or fix carbohydrates. Most lipids are blackened by osmium tetroxide, as it combines chemically to the lipid. The osmium tetroxide must be washed out of the tissue with running water to prevent it from becoming an insoluble precipitate during dehydration.

Osmium tetroxide is mainly used for fixation of tissue for electron microscopy, as this fixing reagent gives a very accurate representation of the ultrastructure of the cell and its inclusions. It is also used for demonstrating lipids as a histochemical technique.

NOTE: Caution should be taken in handling the osmium tetroxide, as its vapors are extremely irritating and destructive to the eyes, mouth, and nose linings.

c. **"Indifferent" or Nonfixative Chemical Reagents.**—These are salts which, when combined with the other fixing reagents, improve the results obtained. Although the mechanism of how these salts act is unknown, it is thought that since the saline solutions used are isotonic to that of the fluids in the tissue, shrinkage is reduced in the presence of these indifferent salts.

3. **Fixatives or Fixative Solutions and Their Uses.**—Fixatives are a mixture of several fixing agents. The fixatives can be classified into two main groups.

a. Routine Fixatives.—These fixatives preserve the relation of specific tissues and cells to each other. They usually consist of coagulant and noncoagulant fixing reagents plus acetic acid, except for formalin fixatives.

b. Special or Cytological Fixatives.—These fixatives preserve the intracellular or cytoplasmic constituents of the cells. They usually consist of coagulant and noncoagulant fixing reagents and more recently of noncoagulant fixing reagents only.

a. **Routine Fixatives.**—There are many routine fixatives. Only a few will be listed to illustrate the combination of coagulant and noncoagulant fixing reagents.

(1) *10% Neutral Buffered Formalin.*—

10% Formalin	1000.0 ml
Sodium dihydrogen phosphate, monohydrate ($NaH_2PO_4 \cdot H_2O$)	4.0 g
Sodium monohydrogen phosphate anhydrous (Na_2HPO_4)	6.5 g

Wash in water or alcohol.

Tissue may remain in the formalin indefinitely, as fixing action is progressive, without affecting the quality of the tissue. Paraffin infiltration damages the tissue since the proteins are not coagulated and stabilized to withstand the rigors of this step. Besides neutral buffered formalin there are many other formulae for formalin fixatives, some containing water, others with salt (sodium or calcium chloride), and some containing alcohol.

(2) *Bouin's Fixative.*—This fixative contains glacial acetic acid to counteract the shrinkage caused by picric acid

1% Saturated aqueous picric acid	75.0 ml
10% Formalin	25.0 ml
Glacial acetic acid	5.0 ml

Fix for 6–24 hr. Wash in 50% alcohol to remove yellow color of the picric acid. Due to the shrinkage caused by the picric acid, large spaces (artifacts) often form in the tissue.

(3) *Zenker's Fixative.*—Zenker's fixative is considered the most efficient fixative for routine tissue preservation and for later staining, as its reagents serve as mordants for the stains. This fixative contains two protein coagulants, acidified potassium dichromate and mercuric chloride, and acetic acid that opposes shrinkage. It does not contain a noncoagulant reagent, and sodium sulfate is the indifferent salt.

Potassium dichromate	2.5 g
Mercuric chloride	5.0 g
Sodium sulfate	1.0 g
Glacial acetic acid	5.0 ml
Distilled water	100.0 ml

It is no longer necessary to add the glacial acetic acid just before use. The C.P. grade glacial acetic acid is free of acetone and alcohol; these reagents are known to reduce the dichromate to chromic oxide, which precipitates as green crystals.

Since the penetration rates of the reagents of this fixative differ, small pieces of tissue, 3–4 mm thick, should be used, otherwise there is poor fixation in the center of large pieces.

Tissue is fixed for 4–24 hr and then washed in running water overnight. Penetration rate of fixative, in general, is rapid.

Postfixation Treatments.—

(a) Post-chromatization.—This is an optional treatment if tissue seems to harden excessively. Following 18 hr in Zenker's fixative, place tissue in a 3% potassium dichromate solution for 12–24 hr. Then wash thoroughly in running water overnight.

(b) Removal of Mercury Pigments.—When tissue has been fixed in fixatives containing mercuric chloride, the tissue must be post-treated with an iodine-thiosulfate sequence to remove the black mercury precipitate and to decolorize the iodine. The treatment can occur during the dehydration or just prior to staining. During dehydration the tissue is placed into 70% alcohol containing an alcoholic Lugol's iodine solution for 5–8 hr. In practice, it is much easier to remove the mercury and decolorize the iodine with the thiosulfate just before tissue sections are stained.

The iodine unites with the mercury to form mercuric iodine, which is soluble in alcohol. The brown color of the tissue sections, due to the iodine, is removed by decolorizing the iodine with sodium thiosulfate.

Alcoholic Lugol's Iodine Solution (Weigert's Variation)

Iodine crystals	1.0 g
Potassium iodide	2.0 g
70% Ethyl alcohol	100.0 ml

Dissolve potassium iodide first in a few milliliters of the 70% alcohol, then the iodine crystals will go readily into solution. Add remaining alcohol to make up to volume (qs).

5% Sodium Thiosulfate

Sodium thiosulfate	5.0 g
Distilled water	100.0 ml

For method of treating tissue sections for removal of mercury pigment and decolorizing the iodine, see Exercise 7.

(4) *Helly's Fixative.—*It contains coagulant (mercuric chloride) and non-coagulant (formalin) fixing reagents, in addition to a fixing reagent of lipids (unacidified potassium dichromate) and an indifferent salt.

Potassium dichromate	2.5 g
Mercuric chloride	5.0 g
Sodium sulfate	1.0 g
Distilled water	100.0 ml
Concentrated formalin	5.0 ml

Immediately before use add the 5.0 ml of concentrated (100%) formalin. Fix for 6–24 hr; post-chromatize tissue if tissue hardens excessively as described above. Wash in running water overnight and post-treat for removal of mercury pigment.

Helly's fixative does not harden the tissue to the same extent as Zenker's fixative, but Zenker's fixative penetrates the tissue more rapidly. Although Helly's and Zenker's fixatives appear to have similar components, the formalin and unacidified potassium dichromate have different effects on the tissue constituents than acetic acid and acidified potassium dichromate. This fixative is excellent as a routine or special cytoplasmic fixative, as it preserves cytoplasmic components. The combination of mercuric chloride and formalin fixes the proteins of the cell's cytoplasm with a meshwork which permits easy infiltration by paraffin and thus tissue has an excellent consistency for sectioning. This fixative is superior, then, for bone marrow, for blood-forming organs such as spleen, for lymph nodes, and for endocrine organs, where preservation of the cells is desired.

(5) *Heidenhain's Susa Fixative.*—This is a general purpose fixative and is a good substitute for Zenker's fixative if potassium dichromate is not needed.

Saturated aqueous mercuric chloride (7%) solution	80.0 ml
Glacial acetic acid	4.0 ml
Concentrated formalin	20.0 ml
Trichloroacetic acid	2.0 g
Sodium chloride	0.5 g

Fix for 3–24 hr depending on thickness of tissue (limit 1 cm thick). This fixative penetrates rapidly and only slightly hardens tissue. Transfer to 70 or 95% alcohol. Post-treat tissue for removal of mercury pigment prior to staining of tissue sections.

b. **Special or Cytological Fixatives.**—

(1) *True Carnoy's Fixative.*—This fixative works well on nerve tissue, especially for subsequent staining of the Nissl bodies. It also fixes glycogen, but dissolves the other cytoplasmic inclusions.

Fixation occurs as rapidly as it penetrates the tissue and dehydration takes place simultaneously.

Glacial acetic acid	10.0 ml
Chloroform	30.0 ml
Absolute ethyl alcohol	60.0 ml

Fix 20 min to 3 hr, as fixative causes shrinkage of the tissue if tissue remains in the fixative for a longer period of time.

The tissue should be no more than 5 mm thick.

Transfer to absolute ethanol to wash out the chloroform, as dehydration has been initiated during fixation.

A variation of this formulation of Carnoy's fixative is the addition of enough mercuric chloride to make a saturated solution in the mixture.

(2) *Lavdowsky Fixative.*—This fixative preserves glycogen.

2% Aqueous chromic acid	10.0 ml
Glacial acetic acid	0.5 ml
95% Ethyl alcohol	10.0 ml
Distilled water	80.0 ml

Fix for 12–24 hr and transfer to 80% alcohol.

(3) *Regaud Fixative.*—This fixative is recommended for the preservation of mitochondria and cytoplasmic granules.

3% Aqueous potassium dichromate	80.0 ml
Concentrated formalin	20.0 ml

Mix reagents just before use.

Fix tissue 4–24 hr. Tissue tends to harden in this fixative; thus post-chromatize tissue several days in 3% potassium dichromate; change solution every day.

Wash tissue 24 hr in running tap water.

(4) *Osmium Tetroxide Fixatives.*—

(a) Altman's Fixative.—This preserves lipids and mitochondria.

5% Potassium dichromate	10.0 ml
2% Osmium tetroxide	10.0 ml

The osmium tetroxide blackens the lipids. Wash in running tap water overnight.

(b) Flemming Fixative (Strong).—This is a classical cytological fixative for preservation of chromosomes so mitotic figures can be demonstrated.

2% Osmium tetroxide	20.0 ml
1% Aqueous chromium trioxide	75.0 ml
Glacial acetic acid	5.0 ml

It is not necessary to use such large volumes since osmium tetroxide has a high cost. Mix reagents just before use.

Fix 12–48 hr. Fixative penetrates slowly, so use 2 mm-thick pieces of tissue. Tissue hardens excessively.

Wash in running tap water for 24 hr. The osmium tetroxide blackens tissue and hematoxylin does not stain the nucleus readily after this fixation. Use safranin stain instead. Due to differential rates of penetration of the components of the fixative the intermediate region of the tissue fixes best.

More recently osmium tetroxide fixatives have been used mainly to prepare tissue for electron microscopy and, as noted before, for the histochemical demonstration of lipids.

c. **Decalcifying Fluids.**—After fixation in a formalin fixative, the decalcifying fluid will soften calcified tissue by removal of the calcium. There are several decalcifying fluids; only one will be listed.

EDTA, disodium salt	5.0 g
10% Formalin	100.0 ml

The EDTA binds or chelates the calcium.

Decalcify 3–4 weeks; change fluid at end of each week.

Tissue preservation is excellent in this decalcifying fluid, although the tissue hardens.

d. **Secondary Fixation.**—Tissues treated with a secondary fixative show general improvement in preservation of tissue components and staining. The tissue is usually first fixed in 10% buffered formalin and secondarily fixed in a coagulant type fixative, such as Helly's fixative. It is not necessary to deformalinize before secondary fixation can occur.

e. **General Properties of a Fixative.**—To fix a tissue properly and to prevent postmortem changes, a fixative must have the following properties:

(1) Must not excessively shrink or swell a tissue.

(2) Must not distort or dissolve tissue components.

(3) Must harden the tissue sufficiently so the tissue retains its form, and its components retain their spatial relationships to each other during subsequent histological steps.

(4) Must inactivate enzymes.

(5) Must kill microbes.

(6) Must improve optical differentiation and enhance staining properties of the tissue components.

f. **Choice of a Fixative.**—Choice of fixative depends on future use of the tissue.

(1) *Routine Versus Special Fixatives.*—Routine or general purpose fixatives will preserve general features; but one must use special or cytological fixatives to preserve specific parts, especially cells, organelles and inclusions.

(2) *Density of Tissue and Rate of Penetration of the Fixative.*—Dense tissues do not fix well in slowly penetrating fixatives; one should use rapidly penetrating fixatives to keep postmortem changes to a minimum.

(3) *Time of Fixation.*—Several factors determine the time tissue is immersed in the fixative. These are action of fixative, rate of penetration, and tissue thickness.

Fixatives containing coagulant reagents, such as Bouin's fixative, fix as fast as they penetrate, i.e., the tissue is fixed as soon as the tissue comes into contact with the fixative. Those fixatives containing noncoagulant reagents, such as 10% formalin, show a progressive improvement in fixing action after the tissue has been completely penetrated by the fixative, i.e., the longer the time tissue is in fixative, the better the tissue is fixed.

(4) *Hardening Effects of the Fixative.*—If the fixative hardens the tissue too much, the tissue will be friable or crumble on sectioning. If the tissue is not hardened sufficiently, the tissue will be too soft and of poor consistency for sectioning.

III. WASHING THE TISSUE

After fixation for the correct number of hours, the excess fixative must be washed out of the tissue to prevent interference with subsequent processing of the tissue. Washing is usually accomplished in cold running tap water 6–24 hr or overnight. Sometimes the tissue may be transferred to alcohol as indicated in some fixation procedures.

A convenient stopping place in the histological procedure is to bring the tissues, after washing in water, through 50% alcohol and then to 70 or 80% alcohol. At these alcohol concentrations, tissue can be stored for several weeks or months without harm to the tissue. Storage for a year or longer reduces stainability of the tissues. Thus it is best to get the tissues embedded in the paraffin as quickly as possible.

QUESTIONS

Cover answers with a piece of paper. Answers appear at end of questions.

(1) When tissue specimens are removed from an animal, they should be immediately placed in
   (a) Clearant
   (b) Dehydrant
   (c) Fixative
   (d) Desiccant

(2) Providing a linkage between tissue components and the stain is referred to as
   (a) Mordanting
   (b) Hardening
   (c) Penetrating
   (d) Fixating

(3) Which of the following is a disadvantage of coagulant fixing reagents?
   (a) Changes fine fibrous network of proteins into coarse network
   (b) Strengthening of chemical bonds
   (c) Permits paraffin to easily infiltrate interior of tissue
   (d) Induces formation of artifacts

(4) Of the following, which is a coagulant type of fixative?
   (a) Chromium trioxide
   (b) Formaldehyde
   (c) Osmium tetroxide
   (d) Potassium dichromate

(5) One of the major problems with picric acid is
   (a) It coagulates proteins
   (b) In its dry state it may explode
   (c) It does not dissolve lipids
   (d) It is purple in color

(6) Formalin (100%) is actually
   (a) 10% Formaldehyde
   (b) 40% Formaldehyde
   (c) 100% Formaldehyde
   (d) 0% since it is a gas

(7) Osmium tetroxide stains__________ black
   (a) Carbohydrates
   (b) Proteins
   (c) All cytoplasm
   (d) Lipids

(8) The most efficient fixative for routine tissue preservation is
   (a) Formalin
   (b) 0.9% Saline
   (c) Picric acid
   (d) Zenker's

(9) For fixation of nerve tissue, especially for subsequent staining of the Nissl bodies, the fixative of choice is
   (a) Zenker's
   (b) Bouin's
   (c) True Carnoy's
   (d) Helly's

(10) Which of the following is not a general property of a good fixative?
   (a) Prevents postmortem change
   (b) Does not distort or dissolve tissue components
   (c) Inactivates enzymes
   (d) Shrinks tissue for easy blocking

Answers

(1) c	(6) b
(2) a	(7) d
(3) d	(8) d
(4) a	(9) c
(5) b	(10) d

3

Dehydration, Clearing, Infiltration, Embedding and Routine Timing Schedule for Manual Technique

Since fixed tissue is not firm enough to section on the microtome, the tissue must be infiltrated and embedded with some supporting substance, such as paraffin. The supporting substance will furnish stability and hold the tissue components in proper relationship to each other. As paraffin and water are not miscible, the tissue must first be dehydrated and cleared, then infiltrated and embedded in paraffin. If these steps of the histological procedure are improperly carried out, especially in terms of insufficient time in each, difficulty in sectioning of the tissue will be experienced.

IV. DEHYDRATION

Tissues fixed in aqueous fixatives have a high water content. The process of removal of this so-called "free" water, i.e., water not bound to the tissue components but within the spaces of the protein network, is termed dehydration. Thorough dehydration and subsequent dealcoholization prior to infiltration and embedding is necessary since in routine processing of tissue nonaqueous infiltration and embedding materials are used.

Routine dehydration in the paraffin histological procedure is accomplished by using a series of gradually increasing percentages of ethyl or isopropyl alcohol, which slowly replaces the water in the tissue, to reduce shrinkage and distortion of tissue components. The dehydration occurs by a process of diffusion due to the inward passage of dehydrating agent (alcohol) and the outward flow of water. When the water and alcohol are at dynamic equilibrium only about 3–4% of the "free" water remains in the tissue.

1. **Graded Series of Alcohols and Length of Time in Each.**—The type of graded alcoholic series and time in each percentage depend on tissue size and type of tissue. The volume of alcohol should be at least 10 times the size of the tissue.

   The percentages of alcohol usually employed are 70, 80, 95 and 100% (absolute) alcohol, and time in each concentration is usually 1 hr, except for 100% where the tissue is placed into two or three changes for 1 hr each. This schedule ensures thorough dehydration. For special, delicate or large pieces of tissue, dehydration should start at either 30% or 50% and might remain for longer than 1 hr in each concentration. If tissue were placed directly into absolute alcohol, a marked turbulence would result caused by rapid mixing of the water and the alcohol and this could cause extreme shrinkage and structural distortion, especially at the surface of the tissue; thus the use of a gradually increasing series of alcohols. In automatic tissue processing, such as in the Ultra-Autotechnicon, a series of gradually increasing concentrations of alcohol is not used, so that in the process you are probably compromising speed of processing the tissue for some artifact, shrinkage. In manual technique where there is a series of beakers containing the alcohols, the tissues are transferred from lowest to highest concentrations at intervals of time. The concentrations of the alcohols become diluted as the tissues pass through. The contamination can be minimized somewhat by allowing tissues to drain and then blotting excess fluid on paper toweling. To effect economy, the alcohol should be "moved down" one, i.e., discard alcohol in the first beaker, replace it with fluid from second beaker and then move the rest of the series down one. The last beaker is replaced with fresh absolute alcohol which should ensure thorough dehydration. The frequency with which this is effected depends on the tissue load going through the series.

To accelerate diffusion by bringing fresh alcohol into contact with the tissue and to hasten dehydration, the alcohol surrounding the tissue should be in constant motion.

Tissue should not remain excessively long in low concentrations of alcohol (under 70%) as these concentrations tend to make the tissue fall apart. Nor should tissue remain too long in high concentrations of alcohol (above 80%) as tissue becomes hard, brittle and crumbly and is then difficult to section. Furthermore, storage of tissue in 70–80% alcohol for a long period of time interferes with staining of the tissue.

2. Dilution of Alcohols.—

a. **Ethyl Alcohol.**—To make various percentages of alcohol, dilute the higher percentage alcohol with distilled water. Ninety-five percent ethyl alcohol is the usual percentage from which the lower concentrations of alcohols are made. Since absolute alcohol is taxed and expensive, the dilutions are not made from 100% alcohol. To make a lower percentage alcohol from 95% alcohol subtract the percentage required from the percentage alcohol (95%) that is to be diluted. The difference will be the amount of water that has to be added.

For example, to make a 70% alcohol solution from 95% alcohol, subtract 70 from 95, 95 − 70 = 25. Thus take 70 ml of 95% alcohol and add 25 ml of distilled water. This will make 95 ml of 70% alcohol (see Fig. 3.1).

Another way to calculate percentage of alcohol is to use the equation

$$V_1 \times \%_1 = V_2 \times \%_2$$

$$V_1 \times 95\% = 95 \text{ ml} \times 70\%$$

$$V_1 = \frac{95 \times 70}{95}$$

$$V_1 = 70 \text{ ml of } 95\% \text{ alcohol}$$

+ 25 ml of distilled water

95 ml of 70% alcohol

V_1 = volume of 95% alcohol to use
$\%_1$ = 95% alcohol
V_2 = ml of diluted alcohol desired
$\%_2$ = desired alcohol percentage

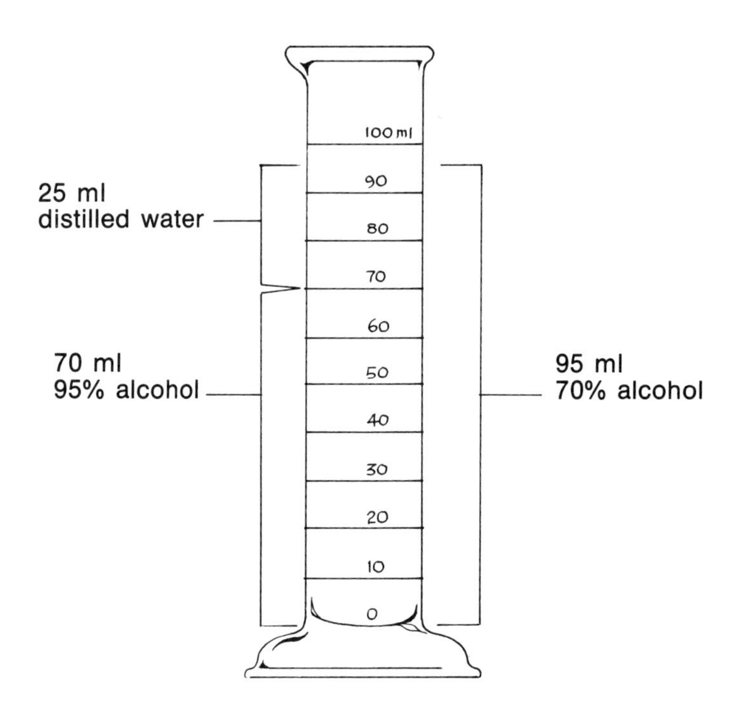

FIG. 3.1. METHOD OF DILUTION OF 95% ALCOHOL TO 70% ALCOHOL

In essence, then, in making the dilution, you are adding a sufficient quantity of distilled water so that the final volume of the diluted alcohol is equivalent to the original concentration of the alcohol from which the dilution is made.

b. **Isopropyl Alcohol.**—Dilutions of isopropyl alcohol are made from absolute isopropyl alcohol.

3. **Types of Dehydrating Agents or Dehydrants.**—

a. **Alcohol.**—Ethyl and isopropyl alcohols are predominantly used in routine dehydration of tissue as they are nontoxic, reliable and fast-acting. They do cause shrinkage and in higher concentrations cause hardening of the tissue.

Butyl alcohol dehydrates slowly and causes less shrinkage and hardening than does ethyl alcohol, but tissue must be kept in each concentration of this alcohol for a long time. Butyl alcohol may be substituted for higher ethyl alcohol concentrations in the dehydration series to prevent overhardening of the tissue.

b. **Dioxane (Diethylene Dioxide).**—The substance can be used to dehydrate and clear as it is miscible with both water and paraffin. Dioxane does not affect tissue texture or staining properties and produces less shrinkage than alcoholic dehydrants. *NOTE:* Its fumes are toxic and it should be used in a well ventilated room! It can also explode with prolonged storage!

c. **Cellosolve (Ethylene Glycol Monoethyl Ether).**—This reagent dehydrates rapidly. Tissue is transferred to several changes of concentrated cellosolve without tissue distortion, shrinkage or hardening. Tissue may be stored several months in cellosolve. The disadvantage of cellosolve is that it appears to dissolve intracellular components even when they are appropriately fixed.

d. **Others.**—Other dehydrants are reagent alcohol (90% ethyl alcohol, 5% methyl alcohol, 5% isopropyl alcohol), acetone, and Technicon's S-29 (an isopropyl type of alcohol). Reagent alcohol is used frequently in a dehydrating series and S-29 is used in Technicon's procedure for the automatic processing of tissue.

V. CLEARING (DEALCOHOLIZATION)

This an intermediate step between dehydration and paraffin infiltration, since alcohol, used in the routine histological procedure, and paraffin are immiscible. The tissue has to be passed through a fluid that is freely miscible with both. The purpose of the step is removal of the alcohol and replacement with a fluid that will mix with paraffin. The tissues are placed in two or three changes of these reagents (volume is 10 times the size of the tissue) for about 1 hr each or longer depending on type and size of tissue. Too long in the clearing agent hardens the tissue and the tissue becomes difficult to section. Minimize contamination by draining and blotting tissues on paper toweling.

The frequency of changing the fluids in this step depends on the tissue load passing through. As in dehydration, discard fluid in first beaker, replace it with fluid from second beaker, and add fresh clearing agent to second beaker. Never discard the clearing agents down the drain of the sink as these highly organic substances do not mix with water. The waste clearing agents should be placed in metal waste can receptacles for later disposal.

This step is called clearing because most, but not all, reagents, as they penetrate, make the tissue transparent and also harden it. This is incidental to the replacement of the alcohol. They increase the transparency of the tissue, as they have the same index of refraction as the tissue. Thus when these fluids penetrate into and fill up the spaces between the tissue elements, they make the tissue optically homogeneous and transparent.

Good clearing agents should replace alcohol quickly and not overharden the tissue.

Types of Clearing Agents or Clearants

NOTE: Most clearing agents are highly toxic and inflammable and should be handled with care!

a. **Xylene.**—It is the most commonly used clearing agent today. It is a colorless reagent. Xylene's action is rapid in displacement of absolute alcohol and leaves tissue transparent. Its disadvantage, besides being highly inflammable and toxic, is that upon prolonged exposure to the tissue, especially to soft tissue, such as brain or spleen, it renders them hard and brittle and difficult to section.

If dehydration is incomplete the xylene becomes opaque or milky as the water comes out of the tissue. If this occurs, put the tissue back into absolute alcohol and continue the dehydration.

b. **Benzene.**—It penetrates and clears tissue very rapidly, produces a minimum of shrinkage, and does not leave the tissue hard and brittle. Its disadvantages are:

(1) Its low boiling point causes benzene to evaporate from the tissue leaving air pockets which would prevent paraffin infiltration.

(2) It is toxic and highly inflammable. It is possibly carcinogenic and can cause aplastic anemia.

c. **Toluene.**—This clearing agent has general properties similar to xylene. It does not harden tissue as much as xylene but penetrates less rapidly than xylene or chloroform.

d. **Chloroform.**—Chloroform penetrates slowly compared to xylene, but causes little shrinkage and does not harden the tissue excessively. It is removed by paraffin infiltration more slowly than the previous clearing agents (a, b and c) and does not clear the tissue. Chloroform is not flammable, but is toxic on inhalation of the fumes.

e. **Cedarwood Oil.**—Cedarwood oil is very penetrating, it produces minimal shrinkage, and tissue may be left in it indefinitely without harm. After clearing in this oil, paraffin infiltration is slow and three changes of paraffin are necessary. To speed up infiltration after cedarwood oil infiltration, place tissue in benzene or toluene for about 30 min, as this will replace the cedarwood oil, then place the tissue into the paraffin. Cedarwood oil is a very expensive substance. There are other clearing oils, such as clove oil and oil of wintergreen (methyl salicylate).

f. **Others.**—In addition to the reagents listed, dioxane, amyl acetate, and carbol-xylene can be used. For the automatic tissue processing by Technicon techniques, C-550 or C-650 is used for the Autotechnicon and UC-670 is used for the Ultra-Autotechnicon.

VI. INFILTRATION

After the tissue has been thoroughly permeated with the clearing agent, the tissue is infiltrated with a supporting medium such as paraffin. Paraffin has several properties that allow it to be a suitable infiltrating medium. These are:

1) It is rapidly converted from solid to liquid form on heating.
2) It permeates the tissue in a liquid state.
3) It solidifies relatively quickly on cooling.
4) It becomes fluid on heating to a temperature which will not damage the tissue.
5) Furthermore, when the paraffin solidifies it becomes firm enough to section at room temperature.

It is necessary to impregnate the fixed tissue with paraffin, as the fixed tissue alone is not of suitable consistency for sectioning. Once the liquid paraffin has infiltrated and hardened, it maintains the components of the tissue in proper relation to each other;

otherwise these components would be compressed and distorted during sectioning. Paraffin infiltration is generally used for routine histological technique.

Advantages of Paraffin Infiltration

1. Time of infiltration and subsequent embedding are relatively short for small pieces of tissue.
2. Thin sections can be cut with the rotary microtome and sections will adhere to each other to form a ribbon.
3. Tissue once infiltrated and embedded can be stored in a dry condition indefinitely without damage to the tissue.

Disadvantages of Paraffin Infiltration

1. Distortion of the histology of the tissue due to shrinkage may occur, especially when sections are being attached to glass slides (see following). This is called paraffin artifact. The degree of artifact is dependent upon the fixative used. Formalin by itself produces the most conspicuous artifact, Bouin's or Zenker's fixatives less.
2. Sectioning of paraffin is difficult at high temperatures. Furthermore, when room has low humidity, static electricity causes sections to adhere to the microtome knife. (There are suitable means of dealing with this problem).
3. Time for infiltration of large blocks of tissue is excessive.

In the paraffin infiltration technique, the paraffin should be prefiltered to exclude dust and debris. The solid paraffin pellets at room temperature are heated to render liquid paraffin. The paraffin should be maintained at a temperature 1°–4°C above its melting point. Large quantities of paraffin are melted down and stored for immediate use in constant temperature paraffin dispensers.

Today, the commercial paraffin is not pure hydrocarbon, but a mixture of several kinds of hydrocarbons and plastic which gives the paraffin its consistency, texture and melting point. The paraffin mixture is classified as soft, medium or hard paraffin. Soft paraffin has a melting point range of 45°–50°C, medium paraffin of 50°–55°C, while hard paraffins have a melting point range of 56°–58°C or 60°–68°C. The choice of the grade of paraffin to use depends on hardness of the tissue, section thickness, and temperature at which tissue blocks are to be sectioned. For hard tissues use hard paraffin; for soft tissues use soft paraffin. Use hard paraffin for thin sections (5–7 μ); use soft paraffin for thick sections. The chief factor that determines which paraffin to use is the room temperature at which the tissue is to be sectioned. If temperature is cool, use soft paraffin. If temperature is hot, use hard paraffin. The paraffin usually preferred since it satisfies the criteria for routine use is the 56°–58°C melting point paraffin.

In the manual technique, infiltration proceeds by placing the tissue in at least two or three changes of liquid paraffin in the paraffin oven (58°–60°C) for a total of 2–4 hr. This will ensure that the tissue is free of clearing agent and is fully permeated with the paraffin. If traces of clearing agent are left in the tissue it prevents paraffin from hardening properly and subsequently sectioning is difficult.

Duration of infiltration depends on size, thickness, density and type of tissue.

Infiltration for an excessive length of time will cause shrinkage and hardening and make tissue friable, i.e., crumble on sectioning. The time for the infiltration step should be "minimum" but "consistent with thorough impregnation"[1] of paraffin.

Muscle and connective tissue should not infiltrate for more than 3 hr, as they become overhardened in the paraffin. Nerve tissue (spinal cord and brain) and skin, as they are dense tissues, infiltrate slowly with paraffin and require 4–6 hr for this step. On the whole, two changes of paraffin of 1 1/2 hr each are sufficient for proper infiltration. Times longer than those suggested and temperature more than 1°–2°C above melting point of the paraffin result in "cooking" of the tissue. The tissue becomes hard and friable and unable to be sectioned. The tissue is removed from the clearing agent, drained, and placed in the first paraffin beaker in the paraffin oven. The volume of paraffin should be 15–20 times the volume of the tissue. It is preferable to leave the beakers in the oven and transfer tissue to them, rather than place cool beakers with the tissue into the oven as time is required to melt the cool paraffin.

If two changes of paraffin are used, paraffin in beaker one, after use, should be discarded as it becomes contaminated with clearing agent. The paraffin in beaker two can be used again as beaker one and the empty beaker filled with fresh paraffin to become beaker two.

Vacuum infiltration reduces time for the preceding procedure, as removal of clearing agent, air and the infiltration of the liquid paraffin into the tissue occur more quickly and completely if carried out under reduced pressure.

The vacuum infiltration occurs in a vacuum oven to which a vacuum pump is attached, which reduces the pressure to 300–500 mm Hg or 15 lb/in.^{2}

VII. EMBEDDING OR CASTING OR BLOCKING

Upon completion of infiltration, the tissue is embedded in the paraffin. A suitable embedding mold is filled with the molten paraffin, the tissue is placed in it and oriented so it is sectioned in the proper plane. The paraffin is cooled and hardened within and around the tissue, enclosing the tissue in a solid mass.

A variety of molds can be used, such as the classical paper boat, the Diamond embedding box (lead L's and a metal base plate), and cardboard and plastic embedding molds. The type of mold used depends on the technician's preference. The Diamond embedding box is useful as it can be adjusted to produce a mold of desired size to accommodate tissues of different sizes. The plastic embedding molds are the molds of choice today, as they are designed to produce a paraffin block that requires little or no trimming as their edges are parallel to the microtome knife edge. In some cases the mold also serves as a frame for the paraffin block for insertion into the microtome object clamp or "chuck."

The technique for embedding of tissue consists of the following steps:

1. Remove the tissue from the tissue capsule and transfer the tissue with warm forceps to a small container of freshly melted paraffin. The tips of forceps are heated in an alcohol lamp or in a forceps warmer. The tips should be hot enough so paraffin does

[1]Drury, R.A.B., and Wallington, E.A. 1967. *Carleton's Histological Technique, 4th Edition.* Oxford University Press, London.

not solidify, but not so hot as to cause paraffin to smoke. The forceps should hold the tissue gently without squeezing. The small container of paraffin should be on a hot plate heated to 60°C or in a small table top paraffin oven, such as a Columbia oven.
2. Fill the bottom of the mold with a small amount of paraffin. The depth of the mold should be at least twice the thickness of the tissue.
3. Warm tips of the forceps, pick up tissue, and place into mold so tissue is toward the bottom of the mold and centered, leaving a margin of several millimeters around the tissue. Also orient the tissue for proper plane of sectioning. Manipulation of the tissue in the mold must be quick, so paraffin does not begin to harden.
4. After tissue is in the mold, fill mold entirely with the paraffin. If tissue begins to float as paraffin is being poured, hold tissue down with warm forceps. As the paraffin begins to harden insert a code number label; the label should not go down to the bottom of the paraffin.
5. Allow the surface of the paraffin block to harden, then immerse the mold into a shallow, cool (10°C) water bath for about 10–15 min to hasten solidification of the paraffin. Do not immerse in the water bath until the surface of the paraffin is hardened, as water, being heavier than paraffin, will run into the bottom of the mold and around the tissue.
6. When paraffin is completely hardened, remove it from the mold.

Cooling Temperature

Cooling of paraffin at 10°C rather than at 0°C prevents cracks from appearing in the paraffin block near the tissue. The paraffin will contract too strongly in the line of least resistance, i.e., right through the tissue, if cooled at 0°C. If paraffin is properly cooled, the crystals of paraffin are small and contiguous with each other. The paraffin will appear clear and homogeneous and there is no layering of the paraffin. Paraffin demonstrating these conditions is best for sectioning.

Air in Paraffin

Paraffin contains dissolved air which, when uniformly distributed, will allow the paraffin to appear clear. Pockets of air produce opaque, milky spots which are called "crystallization." Rapid hardening of the surface of the block will trap the air. If air is trapped, cool paraffin from bottom of mold, while keeping surface liquid for a short time. Then allow surface to harden and place mold into the water bath.

If paraffin does not harden properly as seen by cracks or layering of the paraffin, or if paraffin has trapped air bubbles, sectioning will be difficult. To treat these conditions, remelt paraffin, recover the tissue and repeat the embedding process. Paraffin blocks containing the tissue can be stored indefinitely in a cool place or refrigerator.

Embedding can be somewhat automated as with the Tissue-Tek[2] embedding center. At this apparatus, the technician dispenses the liquid paraffin into warm stainless steel Tissue-Tek embedding base molds, and orients the tissue in the mold. Then the plastic embedding ring is placed over the mold and the entire ring-mold combination is filled with paraffin. The paraffin filled mold is transferred to the chilling plate so the paraffin

[2]Ames Co., Inc., Elkhart, Indiana.

hardens. In order to prevent the hardened paraffin from adhering to the base mold on separation of paraffin from the mold, spray mold before embedding with a mold release (teflon) spray. The advantages of the plastic embedding ring are:

1. You can write the code number on the ring; this eliminates the need for an identification label.
2. The ring serves as a frame for the paraffin block to be inserted into the microtome "chuck."
3. Ease of filing the paraffin tissue blocks.

The advantages of the stainless steel base molds are:

1. The base molds are reusable.
2. The molds are of various sizes to accommodate different sized tissue.
3. The molds, thus the blocks, having parallel sides, eliminate the need for trimming the blocks as their top and bottom edges are parallel to the microtome knife edge.

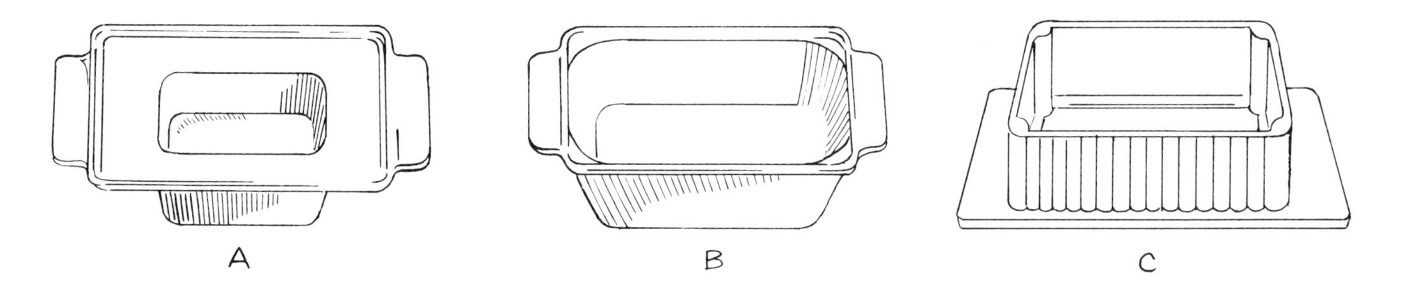

FIG. 3.2. TISSUE-TEK STAINLESS STEEL BASE MOLDS AND PLASTIC EMBEDDING RING
A—and B—Embedding base molds (reusable). C—Embedding ring (disposable).

VIII. ROUTINE TIMING SCHEDULE FOR MANUAL TECHNIQUE

The timing schedule given below is a guide and subject to modification in terms of steps used and time. The schedule depends on type of fixative used and size and nature of tissue being processed. The fixative employed will determine at which point of entry the tissue will enter the schedule and what postfixation treatment is necessary. Size and nature of tissue will determine the time the tissue is in each step of the processing schedule.

1. **Fixation.**—12–24 hr. Consult specific fixation recipes for time of fixation and for any postfixation treatment required.

2. **Wash.**—Use running tap water, if necessary, 6–8 hr or overnight.

3. Dehydration.—

a. 70% Alcohol (either ethyl or isopropyl alcohol)	1 hr
b. 80% Alcohol	1 hr
c. 95% Alcohol	1 hr
d. 100% Alcohol-I	1 hr
e. 100% Alcohol-II	1 hr

4. Clearing.—

a. Clearing agent-I (xylene, benzene, etc.)	1 hr
b. Clearing agent-II	1 hr

5. Infiltration (in Paraffin Oven).—

a. Paraffin-I (56°–58°C)	1 1/2 hr
b. Paraffin-II	1 1/2 hr

6. Embed

The usual method for processing tissue is to pass the tissue from one fluid to another, draining the tissue between steps to avoid carry-over contamination. This method leads to excessive manipulation of tissue with the forceps, which might damage the tissue on handling. To avoid handling of the tissue with forceps use the method of decantation. In this method the tissue is in one beaker only and the fluids are added to and poured off from the beaker. The beaker is drained of the fluid as much as possible before the next solution is poured in. This reduces handling of the tissue and injury to the tissue.

QUESTIONS

Cover answers with a piece of paper. Answers appear at end of questions.

(1) The purpose of paraffin infiltration is to ensure
(a) Dehydration
(b) Removal of xylene
(c) Make tissue firm enough to section
(d) Mix the "free" water with paraffin

(2) Dehydration of tissues for paraffin infiltration involves
(a) A graded series of alcohols
(b) A graded series of alcohols or dioxane
(c) Cellosolve
(d) (a), (b) or (c) may be used

(3) A major characteristic of a good clearing agent is that
(a) It must be freely miscible with the dehydrant and paraffin
(b) It should be inexpensive
(c) It should never need replacing
(d) It should not be toxic or inflammable

(4) Which of the following is not a property that makes paraffin a good supporting medium
(a) It is rapidly converted from solid to liquid
(b) It permeates tissue when in a liquid state
(c) It remains soft at room temperatures
(d) It can be obtained in varying temperature grades

(5) Which of the following is not a factor in the time required for infiltration
(a) Size of piece of tissue
(b) Thickness of tissue
(c) Density and type of tissue
(d) Melting point of paraffin

(6) On the whole, which of the following should be adequate for paraffin infiltration?
(a) Two changes of 1 1/2 hr each
(b) Three changes of 1 hr each
(c) Four changes of 1 hr each
(d) Immersion in one container for 3 1/2 hr at 5°C above melting point

(7) Fixation time should usually be
(a) 2 hr
(b) 3 hr
(c) 12–24 hr depending upon fixative used
(d) 1 1/2 hr in paraffin #1

(8) In the following dehydration schedule, pick the letter of the first step that is out of proper sequence
(a) 70% Alcohol—1 hr
(b) 80% Alcohol—1 hr
(c) 100% Alcohol—1 hr
(d) 95% Alcohol—1 hr

(9) An advantage to the Diamond embedding box is that
(a) It is disposable
(b) It can be adjusted to produce a mold of different sizes
(c) It is made of plastic
(d) It requires absolutely no trimming

(10) Which of the following is incorrect regarding melting points of paraffin
(a) Soft paraffin 45°–50°C
(b) Medium paraffin 50°–55°C
(c) Hard paraffin 56°–58°C
(d) Hard paraffin 68°–72°C

Answers

(1) c
(2) d
(3) a
(4) c
(5) d
(6) a
(7) c
(8) c
(9) b
(10) d

4

Automatic Tissue Processing

Tissue can be rapidly processed without any attention on the part of the technician from fixation through infiltration by automatic tissue processors, such as Autotechnicon[1] or the Autotechnicon Ultra[1]. The automatic tissue processors are invaluable aids in the pathology laboratory where large volumes of tissue have to be processed.

I. PRINCIPLES OF OPERATION

The most recent automatic tissue processors use vacuum, vertical oscillation, and heat to rapidly promote chemical activity, penetration and interchange of the various solutions used in processing of the tissues. The vacuum also eliminates the possibility of trapped air bubbles in the tissues, which would retard penetration of solutions.

The older Autotechnicon models use the same principles as the Ultra except that the tissue is not processed under vacuum and heat; only vertical oscillation is used to agitate the tissues.

A. Heating Bath

A series of plastic beakers containing the various solutions are arranged around a circular deck. The beakers are mounted in a controlled temperature heating bath. The Ultra I contains heated circulating mineral oil, the Ultra II forced hot air. The heating bath is sectioned into two heat zones. The processing section, zone 1, contains ten beakers in which there are successively, fixative, dehydrant and clearing agent; maintained at a temperature between 43° and 45°C. The paraffin section, zone 2, contains two beakers of paraffin and maintains the temperature between 58° and 60°C.

B. Timing Device

The tissues are automatically shifted through the series of beakers by a timing device. The timing device consists of a time clock, a timing pin, a notched timing disc, which has a punched out time cycle program, a locked on cutoff pin and a cutoff cam. Small pieces of

[1]Technicon Instruments Corp., Tarrytown, N.Y.

tissue, 1–3 mm thick, are placed in coded tissue capsules and placed in the tissue basket. While tissue specimens are being taken, the tissue basket is sitting in a beaker of fixative. After all tissues are obtained, the tissue basket is covered and attached to the bottom of the vacuum head.

C. Dome

The vacuum head and beaker covers are suspended from the underside of the dome of the instrument and they rotate with movement of the dome. The beaker covers prevent evaporation and contamination of the solutions, while the vacuum head, to which the tissue basket is attached, allows evacuation of the beaker which the tissues are in. The dome vertically oscillates 10 strokes/min, which brings fresh solution constantly in contact with the tissues. The dome also rotates the tissue basket through the various solutions according to the precise timing program notched on the timing disc.

D. Main Operating Switch, Pilot Light, Heater Indicator Lights, Vacuum Gauge

The main operating switch of the instrument has three positions: the center neutral "off" position, the upper "momentary" position, which when held down allows manual control of the instrument, and the lower "run" position, which when pushed, automatically controls the processing of the tissues. The pilot light glows when power is on to the instrument. There are also indicator lights, which glow when heaters are on in the processing and paraffin sections of the heating bath.

On pushing the operating switch to "run," the dome rotates and the tissue basket descends into the first beaker filled with the fixative. This beaker is evacuated to a negative pressure of 15 lb/in.^{2} (300–500 mm Hg) as indicated on the vacuum pressure gauge, and the tissues are agitated and heated. With the cutoff cam and cutoff pin in position, the timing pin is placed against the timing disc by flipping the locking pin up. This electrically activates the time clock and begins the timing sequence. As the timing disc revolves on the clock, the timing pin riding against the outer edge of the timing disc falls into the first notch of the disc. When the pin enters the notch, it energizes the rotating mechanism of the dome. The vertical oscillation of the dome stops, and the vacuum is released in the beaker as indicated by the pressure gauge returning to zero. The vacuum head and attached tissue basket slowly rise vertically up from the beaker and hover momentarily as the fluid drains off into the beaker. The tissue basket is then shifted and descends into the beaker next in line. This beaker is evacuated and the tissue is again agitated and heated. These same steps are repeated through the eleven remaining beakers. When the tissues are in the last paraffin beaker (beaker 12), the processing cycle is completed. To prevent the tissue from returning to beaker 1, the time clock is shut off as the cutoff pin pushes the cutoff cam against the timing pin. The timing pin is moved away from the timing disc; this stops the clock. The tissues remain in the last paraffin beaker under vacuum, agitation and heat until the technician removes them.

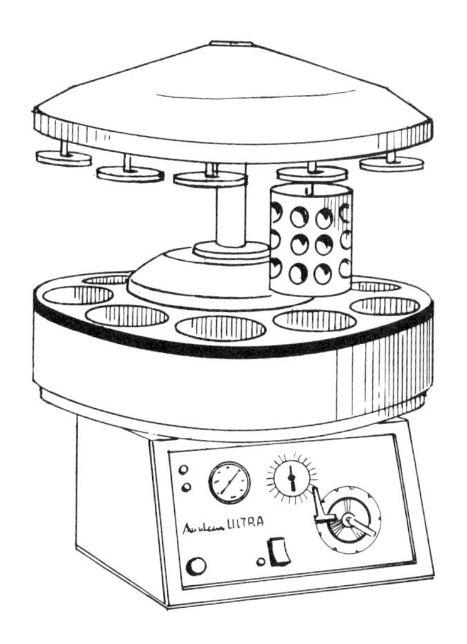

FIG. 4.1. AUTOTECHNICON ULTRA

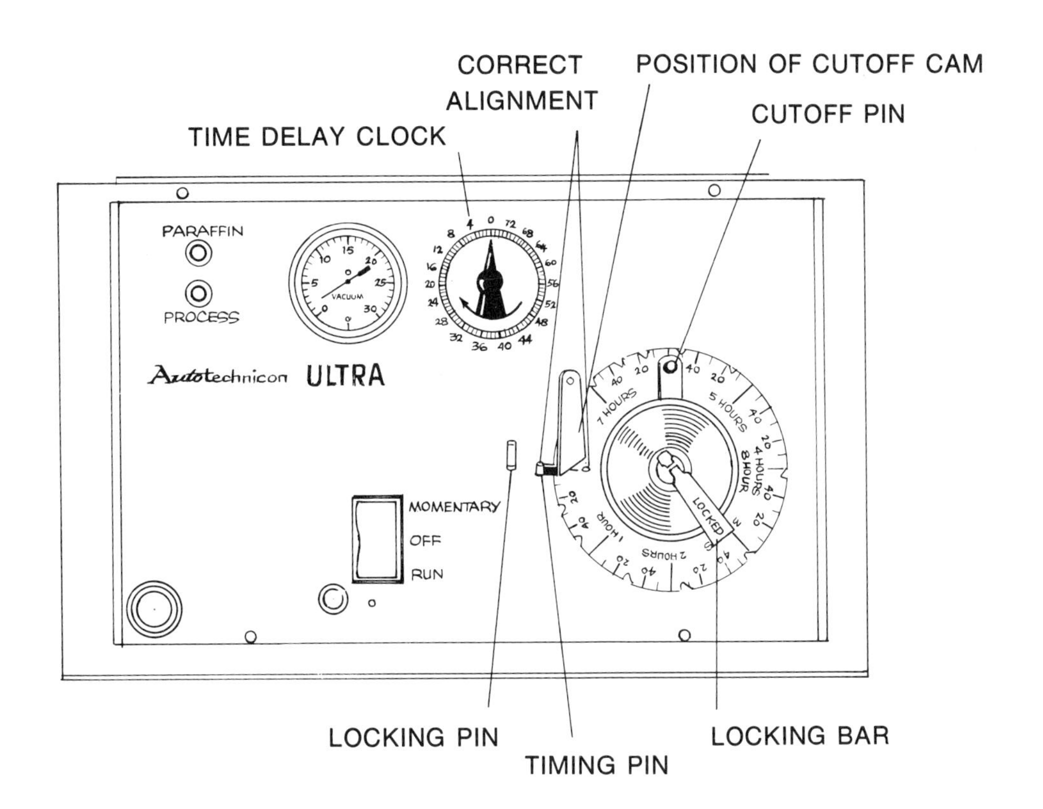

FIG. 4.2. FRONT PANEL INSTRUMENTATION OF AUTOTECHNICON ULTRA

E. Time Delay Clock

The instrument also contains a time delay clock that defers starting of time of processing up to 72 hr by setting the time delay knob to the number of hours you desire to elapse before the processing begins. When the time delay knob reaches zero, the processing of the tissue automatically starts as the timing pin is preset against the timing disc. During the time delay the tissues remain in the fixative.

However, since activation of the time delay clock cuts off power to the heater until the indicated time lapse is completed, this will cause the paraffin to solidify. Since it takes 8 hr of heating to completely liquify the paraffin in beakers 11 and 12, a processing time of 8 hr or shorter can not be used as this will bring tissues into incompletely melted paraffin.

Though unusual, time delay processing may occur over the weekend, when the pathology laboratory may be closed from 5 p.m. Friday evening to 9 a.m. Monday morning, and one wants tissues ready for embedding by this time. For example, if a 16 hr processing cycle is used, just delay the processing for 48 hr to 5 p.m. Sunday, at which time the processing of the tissues will automatically start and the tissues will be ready for embedding at 9 a.m. Monday.

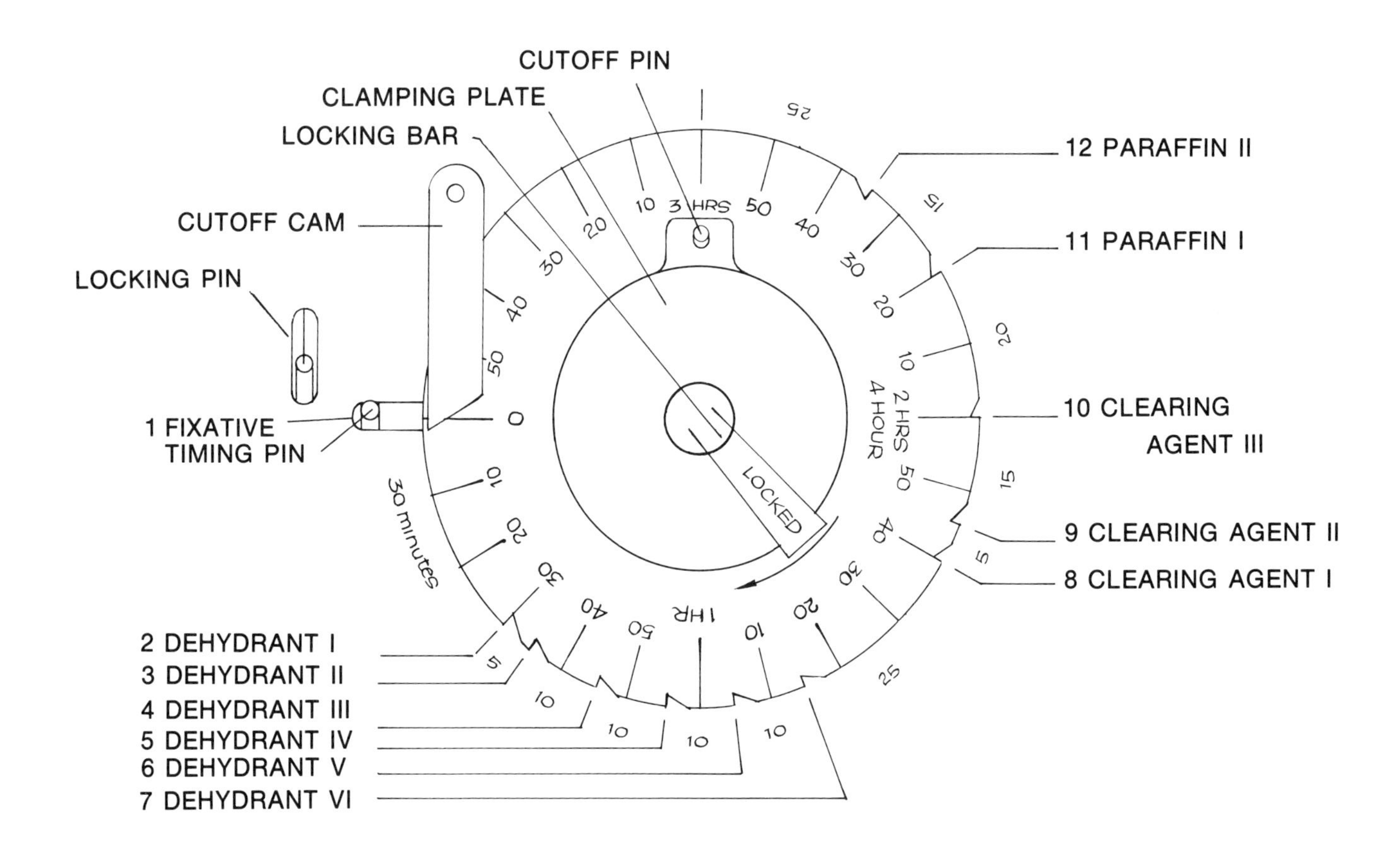

FIG. 4.3. NOTCHED TIMING DISC FOR THE AUTOTECHNICON ULTRA FOR A 3 HR TIMING PROGRAM

Do not notch the zero time or finishing time positions. A correctly notched disc has one notch less than the number of beakers (thus 11 notches).

TABLE 4.1
PROGRAM SCHEDULE FOR A 3 HR TIMING CYCLE

Procedure	Time (min) Amount of Time Tissue Is in Each Beaker	Time Elapsed
		0
1. Fixation	30	
		30
2. Dehydration I	5	
		35
3. Dehydration II	10	
		45
4. Dehydration III	10	
		55
5. Dehydration IV	10	
		65
6. Dehydration V	10	
		75
7. Dehydration VI	25	
		100
8. Clearing I	5	
		105
9. Clearing II	15	
		120
10. Clearing III	20	
		140
11. Infiltration (paraffin I)	15	
		155
12. Infiltration (paraffin II)	25	
		180 end (3 hr)

II. CHEMICAL REAGENTS USED FOR AUTOMATED TISSUE PROCESSING

The chemical reagents used in the Autotechnicon procedure could be similar to those used in manual technique. However, the Technicon Corporation has designed and patented reagents for automated tissue processing.

A. Fixative

FU-48 is a modified Zenker's type fixative, which does not require time consuming washing in water. It is prepared by dissolving the quantity supplied in 10 liters of distilled water.

Another fixative suggested for automated processing is a 1:1 mixture of 10% neutral buffered formalin and the Technicon dehydrant S-29.

B. Dehydrant

S-29 is an isopropyl type of alcohol with additives that prevent it from being hydroscopic and as volatile as regular alcohol. S-29 is used full strength as it does not cause tissue shrinkage and hardening, thus eliminating the time consuming graded series of regular alcohols.

C. Clearing Agents

Technicon's clearing agents have been designed to prevent the tissue from becoming friable and hard. The UC-670 has been especially designed for the Autotechnicon Ultra where processing of the tissues is under heated conditions, but it can be used for all Technicon's processing systems. C-550 and C-650 are used in the older Autotechnicon instruments which do not process tissue under heat or vacuum.

D. Paraffin

Any high grade, prefiltered paraffin can be used such as Tissuemat[2] or Paraplast[3].

III. SCHEDULE FOR CHANGING SOLUTIONS IN THE AUTOTECHNICON ULTRA

The changing of the solutions in the beakers depends on volume of tissue going through the instrument. The usual schedule follows. The fixative is changed daily and the beaker washed out thoroughly. Dehydrant beakers are rotated daily. Discard dehydrant in the first beaker, fill with fresh dehydrant and rotate it to last position; the other beakers should be moved forward one position. Beakers containing clearing agent should be rotated every two or three days in a similar manner. Paraffin is changed daily in the routine as given above.

IV. OTHER AUTOMATIC TISSUE PROCESSORS

Besides the Autotechnicon Ultra I and II, there are other automatic tissue processors, such as the older Autotechnicon Mono and Duo, Fisher Tissuematon, Lipshaw Trimatic and Model 1000. One should consult with manufacturers for methodologies utilized in their instruments.

[2]Fisher Scientific Co., Pittsburg, Pa.
[3]Sherwood Laboratories, Inc., St. Louis, Mo.

QUESTIONS

Cover answers with a piece of paper. Answers appear at end of questions.

(1) Which of the following is not utilized in the latest automatic tissue processing?
(a) Vacuum
(b) Vertical oscillation
(c) Heat
(d) Horizontal inversion

(2) The difference between older Autotechnicon models and the Ultra is that the Ultra
(a) Is mechanically different
(b) Uses vacuum and heat
(c) Requires no clock mechanism
(d) Requires no beakers

(3) In automated tissue processing, ten strokes per minute refers to
(a) Dome oscillation time
(b) Clock mechanism
(c) Timing pin
(d) Notched disc timer

(4) The main operating switch of the Autotechnicon has three positions. Which of the following is not one of them?
(a) Off
(b) Rotate
(c) Momentary
(d) Run

(5) When the "time delay" clock is set, tissue will remain in
(a) 70% Alcohol
(b) Paraffin
(c) Fixative
(d) Air

(6) In a typical 3 hr timing schedule, tissue would be in Dehydration Beaker VI for
(a) 10 min
(b) 30 min
(c) 5 min
(d) 25 min

(7) A good fixative for automated processing is
(a) Bouin's
(b) 1:1 Mixture of 10% neutral buffered formalin and Technicon dehydrant S-29
(c) UC-670
(d) C-550 or C-650

(8) The changing of solutions in the beakers of an automated system is mainly dependent upon
(a) Volume of tissue going through
(b) Cost
(c) Timing setup
(d) Size of tissues

(9) The reason that the tissue basket hovers momentarily before moving to the next position is to allow
(a) Air to penetrate tissues
(b) Air to leave tissues
(c) Partial drying of tissues
(d) Draining of excess fluid

(10) Heating is accomplished in the Ultra II by
(a) Circulating mineral oil
(b) Forced hot air
(c) Chemicals
(d) (a), (b) and (c)

Answers

(1) d	(6) d
(2) b	(7) b
(3) a	(8) a
(4) b	(9) d
(5) c	(10) b

5

Microtomy or Sectioning

I. ROTARY MICROTOME

The rotary microtome is most widely used to cut thin sections of paraffin infiltrated tissues.

A. Principles of Operation of the Rotary Microtome

Here we will discuss the American Optical Corporation's[1] rotary microtome. Many other models exist based on the same principles, for example, Lipshaw.

The microtome has a firm support for the knife and paraffin tissue block and an automatic feed mechanism which is designed to advance the tissue a set number of microns during each cutting stroke. In the rotary microtome the tissue block is mounted in the instrument so that it moves reciprocally up and down in the vertical plane. The tissue block, however, in relation to the cutting edge of the knife, is in the horizontal plane at right angles to the knife, and as it crosses the vertically mounted stationary knife, the sections are cut.

By turning the rotary handle or flywheel of the microtome in a clockwise direction, a complete cutting stroke cycle is produced. On the downward part of the cutting stroke cycle, the tissue block crosses the cutting edge of the knife and a section is cut. On the upward part of the cutting stroke cycle, the microtome's mechanism automatically advances or feeds the tissue into the cutting edge for the next section. The tissue feed mechanism is usually a triangular wedge moving along an inclined plane, which advances the tissue into the cutting edge of the knife. As the sections are cut, the friction of sectioning produces sufficient heat to soften the paraffin and sections adhere one behind the other to form a ribbon. All in all, the microtome is very similar to the way a meat slicer works in the local delicatessen, except the slices on the microtome are extremely thin and the mechanism is highly refined.

B. Microtome Knife

The microtome knife is made of high grade stainless steel. It also has its own fitted back and handle to use when manually sharpening the knife.

[1]American Optical Corp., Buffalo, N.Y.

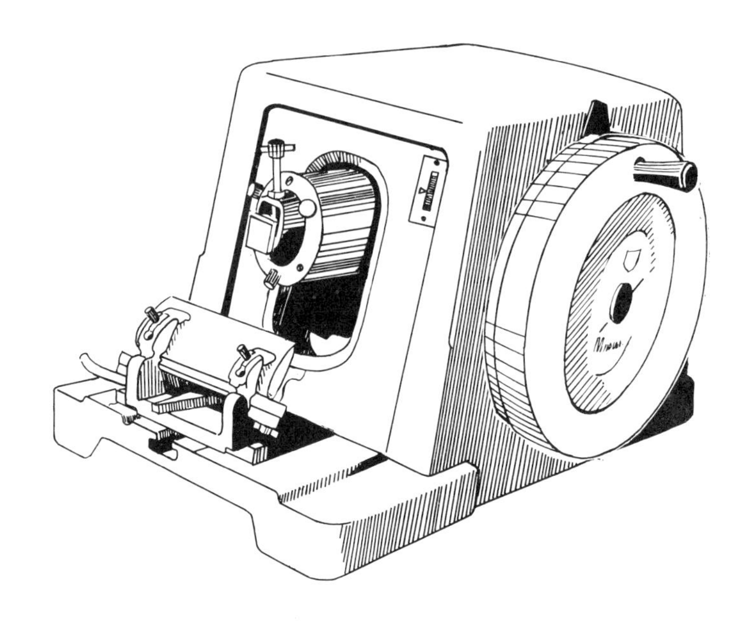

FIG. 5.1. ROTARY MICROTOME

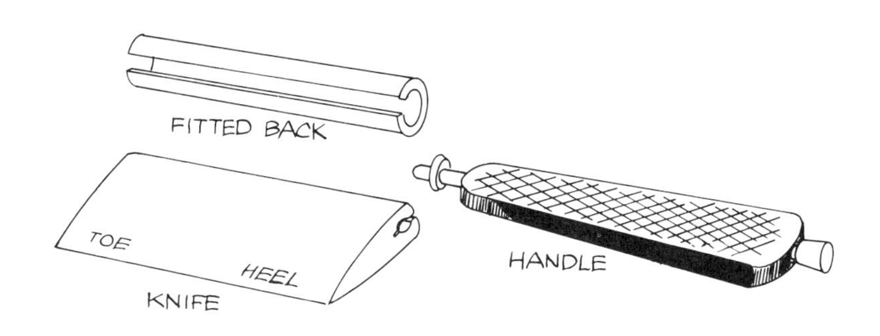

FIG. 5.2. MICROTOME KNIFE WITH FITTED BACK AND HANDLE

1. **Wedge Angle.**—The microtome knife used to cut paraffin embedded tissue on the rotary microtome is wedge shaped with plane or flat surfaces on both sides. The plane surfaces should theoretically meet at an angle of 14°–15°; this is the wedge angle.

2. **Facet or Bevel Angle.**—The actual cutting surfaces of the knife are more inclined toward each other than the plane surfaces of the knife. The actual cutting surfaces of the knife are known cutting facets or bevels. The angle between the cutting facets at the cutting edge of the knife is about 30° and is known as the facet or bevel angle. The width of the cutting facet is 0.3 mm. The smaller the facet angle, the sharper the cutting edge, but the more fragile the knife edge. The greater the facet angle, the less sharp the cutting edge of the knife. The facet angle is maintained during manual sharpening by its fitted back, which raises the knife just enough to form the correct angle between the cutting facets. Each knife has its own back and it should never be interchanged for another.

   The edge of the cutting facet of a sharp knife, when viewed under low power of the microscope by reflected light, appears as a straight, fine line with only slight reflection. Under high power, the edge will have a fine serrated appearance. The closer or finer the serrations, the sharper the cutting edge of the knife. If the cutting edge is dull or nicked, it will reflect more light and, observed along with the nicks, the edge will appear as a coarser serrated line.

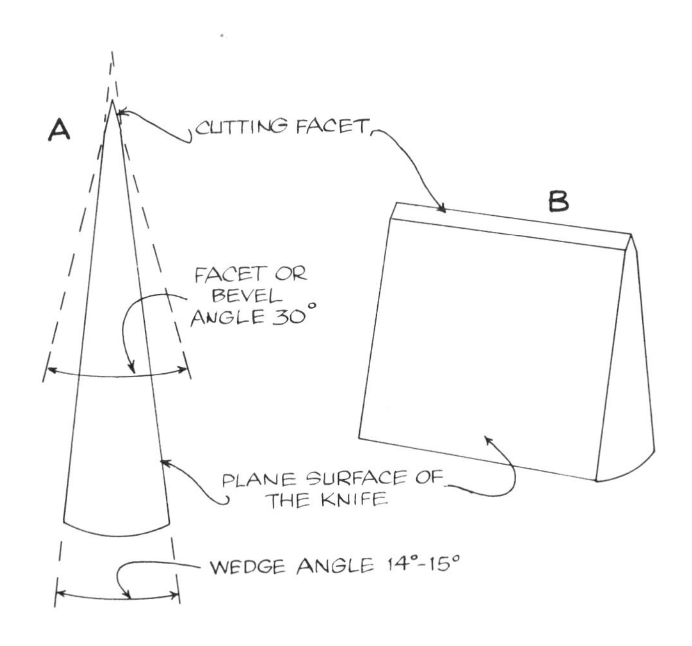

FIG. 5.3. (A) ANGLES OF MICROTOME KNIFE SURFACES
(B) CUTTING FACET

C. Razor Blade and Razor Blade Holder

Single edge super stainless steel razor blades may be substituted for the microtome knife for student use. The Gillette[2] valet blades are good to use, since they have no backs. The removal of the back sometimes leads to cut fingers if one is not careful. The razor blade fits into the razor blade holder (see Fig. 5.4). The razor blade holder uniformly clamps the razor blade securely along its cutting edge with the cutting facets just showing beyond the jaws of the holder. The razor blade holder is then clamped into the microtome knife holder.

Advantages of using razor blades: (1) Sharpening of the blade is not required; blades are discarded when dull. (2) Cutting gritty material with a razor blade is cheaper than dulling and nicking an expensive microtome knife. (3) Cutting edge of razor blade does not extend beyond the ends of the knife holder, thus less chance of accidentally cutting the fingers.

Disadvantages of using razor blades: (1) The razor blades are not as rigid as the microtome knife. (2) The cutting edge of the razor blade is short. (3) If microtome knife and razor blades are used interchangeably, readjustment of angle or tilt of the knife holder is necessary to obtain ribbons of paraffin with each.

[2]American Optical Corp., Buffalo, N.Y.

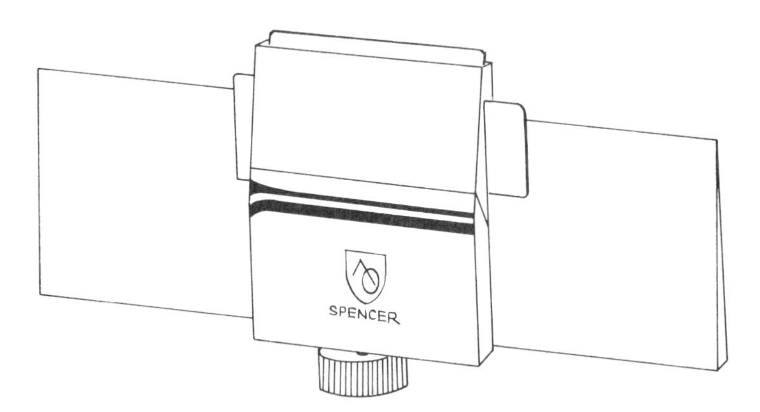

FIG. 5.4. RAZOR BLADE IN HOLDER

D. Sharpening the Microtome Knife

1. **Manual Sharpening.**—The microtome knife is manually sharpened by the method of honing and stropping. Honing or grinding removes nicks and other irregularities of the knife edge. Honing is accomplished with relatively coarse abrasives on a hard stone-like surface. Stropping polishes and sharpens the knife edge. It is carried out on a soft surface with very fine abrasives. During the sharpening procedure, the knife edge should be inspected under the microscope to check on the progress of the honing and stropping.

   a. **Honing.**—Manual honing uses as a hone either a yellow Belgian stone or a white Arkansas stone. The hone is lubricated with soapy water. Coarse hones remove deep nicks, fine hones remove minor imperfections in the knife edge or give a final honing after the coarse hone.

      The initial honing pattern is from heel to toe end of the knife. With the knife edge facing into the hone, the knife is moved away from the technician on a diagonal to the right from bottom left to top right of the hone. The knife is rolled over on its back and returned to the technician on a diagonal to the left from top right to bottom left of the hone. These strokes are repeated the same number of times (see Fig. 5.5). Move the knife to the right, to the toe of the knife, and with the knife edge facing into the hone, the knife is moved away from the technician on a diagonal to the left from bottom right to top left of the hone. The knife is rolled over on its back and returned to the technician on a diagonal to the right from top left to bottom right of the hone. Repeat the same number of strokes as above. Continue coarse and fine honing in the preceding described patterns until the cutting edge irregularities are eliminated by checking under the microscope. As a result of honing, the cutting edge is seen as a series of alternate ridges and grooves due to the grains of abrasive in the stone. Coarse honing produces relatively deep ridges and grooves, while the fine hone reduces their size and brings the knife edge to a semi-polished state.

   b. **Stropping.**—Stropping requires a leather strop with coarse and fine grade sides. The strop is mounted on a revolving wooden block and cushioned by a felt pad. Coarse stropping removes most of the remaining ridges and grooves, i.e., the "rough" edge. Fine stropping polishes the cutting edge, i.e., converts the cutting edge into a straight, smooth line.

When stropping, the knife and strop surfaces should be clean and dry. To maintain the strop use saddle soap to wash the leather and neat's-foot oil to retain the leather's pliability. Work the oil in over small areas at a time; buff with a towel. Do not allow oil to sit on the leather for an extended period of time. Do not use mineral oil. Rubbing a hand over the leather improves its texture.

The stropping patterns are similar to those of honing, except the cutting edge moves away from the strop to prevent the knife edge from cutting into the strop (see Fig. 5.6). Continue stropping with the coarse strop until the "rough" edge is smoothed by checking under the microscope. Then only a few strokes are needed on the fine strop to give a final polish to the cutting edge. Do not overstrop the edge as this causes rounding of the edge and loss of sharpness. Twenty to thirty double strokes should be adequate to polish the knife edge. After stropping the cutting edge, inspection under the microscope should reveal a fine, bright, straight line over its entire length.

During manual sharpening, the right hand usually holds the knife handle. Gentle pressure from the fingers of the left hand against the backing of the knife holds the knife edge evenly to the hone and strop and guides the knife across them. Use the same pressure on the forward and back strokes. Roll the knife on its back without lifting it from the hone and strop. The sharpening motion should be rhythmic at a moderate pace, about one second per stroke.

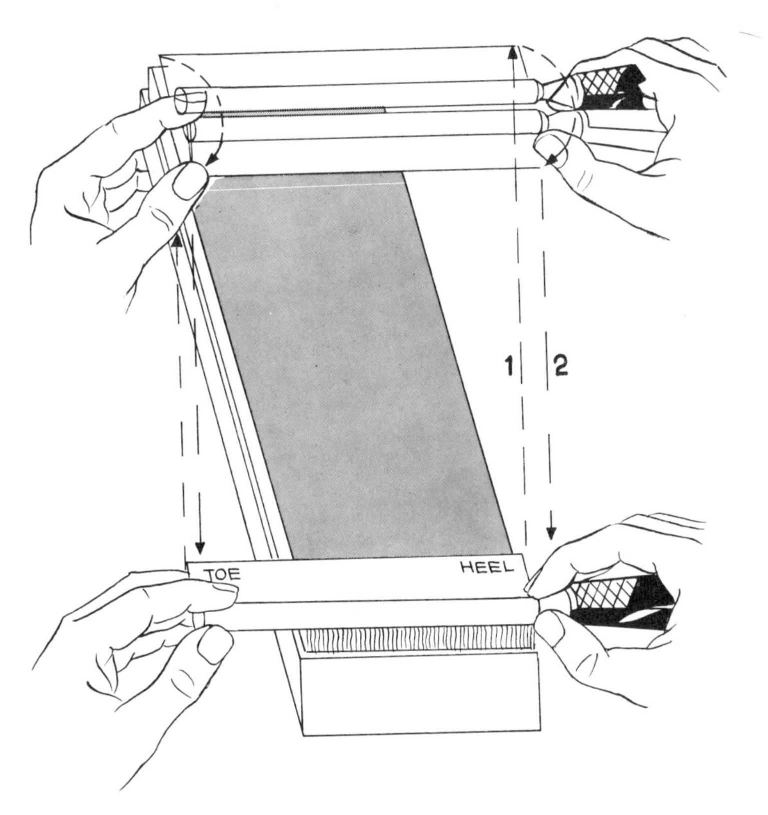

FIG. 5.5. STROKE PATTERN FOR MICROTOME KNIFE HONING

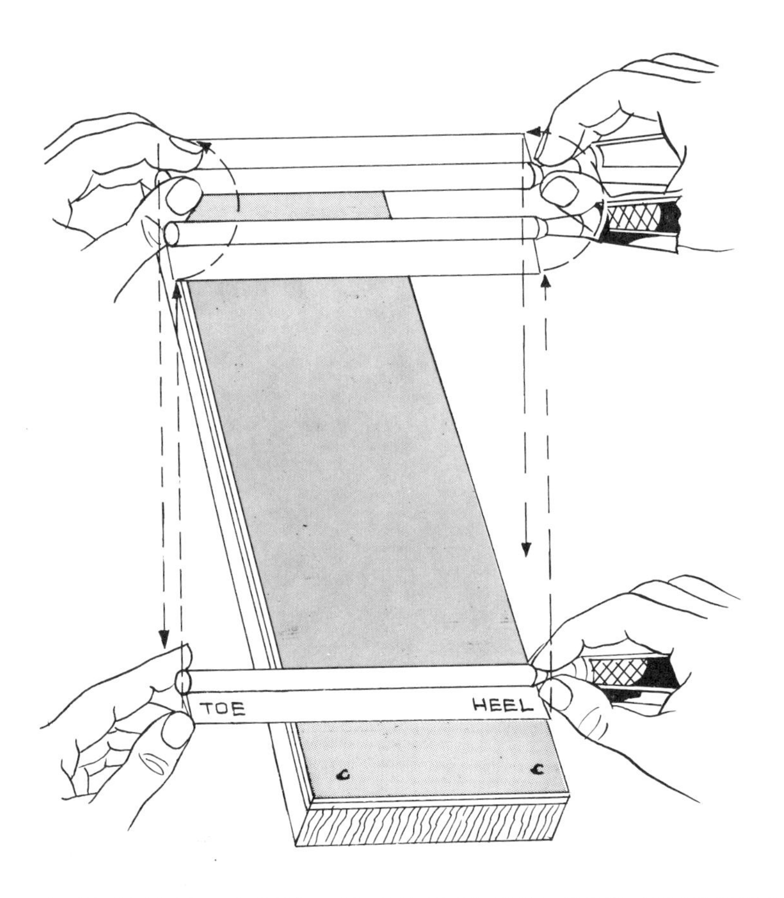

FIG. 5.6. STROKE PATTERN FOR MICROTOME KNIFE STROPPING

2. **Mechanical Sharpening.**—The manual method for knife sharpening has been replaced by mechanical knife sharpeners, which automatically duplicate the manual procedures. Several companies distribute these instruments, such as American Optical Corporation, Scientific Products, Hacker Instruments. Instructions for use accompany the apparatus, so they will not be discussed here.

   For badly worn knives or knives with large nicks, the knife must be reconditioned by a professional knife grinder.

II. PREPARATION OF PARAFFIN TISSUE BLOCK

A. Trimming the Face of the Paraffin Tissue Block

Before inserting the paraffin tissue block into the microtome's chuck, be certain that the upper and lower edges of the block are parallel to each other and to the knife edge, otherwise ribboning of the tissue sections will not occur, or if there is ribboning, a straight ribbon will not be obtained. Curvature of the ribbon will also occur if the lateral sides of the block are not of equal width. Trim the face of the block with a heated scalpel to make these adjustments. Furthermore, if there is excess paraffin on the face of the tissue block, trim the block, leaving just a few millimeters of paraffin surrounding the tissue. An excess of paraffin can cause wrinkling, tissue deformation and prevent flattening of the sections when sections are spread. After these adjustments to the face of the block have been made, attach paraffin block to object carrier as described in Section B following.

B. Attachment of Paraffin Tissue Block to Object Carrier

If the mold that was used for embedding is removed after the paraffin solidifies, the paraffin block must be fused to an object carrier of either metal or wood by paraffin. Usually the face of the object carrier is grooved to increase surface contact between the paraffin block and the tissue carrier.

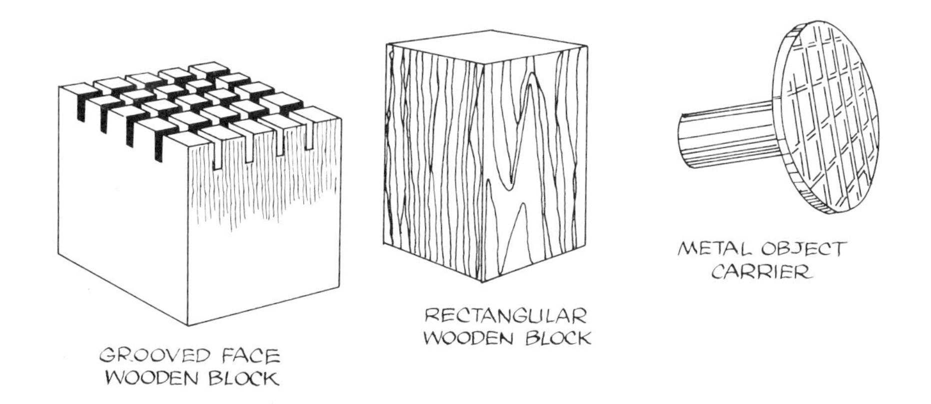

FIG. 5.7. OBJECT CARRIERS USED FOR ATTACHMENT OF PARAFFIN TISSUE BLOCK

One method of attaching the paraffin tissue block to the object carrier is described here. Trim the back end of the paraffin block with a heated scalpel. Heat the shavings to liquid on a metal spatula blade over an alcohol lamp and coat the face of the object carrier with the liquid paraffin. Before the paraffin hardens fuse the paraffin block to the face of the object carrier. The fusion can be accomplished either (1) by briefly passing the back end of the paraffin block through the alcohol flame or Bunsen burner to soften the paraffin, then pressing this end of the paraffin block onto the face of the carrier or (2) by heating the spatula blade and pressing the paraffin block onto the face of the carrier with the heated spatula blade in between, thus heating two surfaces so that when the paraffin is slightly melted, the spatula blade can be withdrawn. The code number is then secured to the tissue block. The attached paraffin block is cooled and hardened by immersion in a shallow pan of cold (10°–15°C) tap water. If the water is too cool, separation of paraffin block from object carrier may result during sectioning.

III. ADJUSTMENTS OF THE ROTARY MICROTOME AND SECTIONING

A. Adjustments of the Rotary Microtome

1. The microtome should be clean before beginning sectioning. Brush paraffin into plastic waste receptacle which is attached below the microtome and wipe all microtome surfaces with a xylene-moistened rag.

2. With rotary handle of the microtome locked by the brake, remove the microtome knife holder. Lift cover of the microtome to check on how far the tissue feed mechanism has advanced. If the tissue feed mechanism has advanced to its maximum; i.e., the triangular wedge is at the extreme upper end of the inclined plane and the gears of the mechanism have disengaged, thus the microtome is in "neutral" position; rewind the crank handle on the side of the microtome until mechanism is fully retracted, i.e., the triangular wedge is at the extreme lower end of the inclined plane, and reengage gears by pulling up on the spring loaded lever which is located behind the feed mechanism. If the gears are not engaged, the feed mechanism will not advance the tissue into the knife edge.

3. Set section thickness as desired. Routine sections are cut between 6 and 10 microns, with average section thickness 6 microns. Some prefer 5 micron sections. Begin sectioning at 6 microns; if sections are difficult to make, raise thickness setting 1 micron at a time, until 10 micron setting is reached.

4. Insert paraffin tissue block into the chuck of the microtome and tighten into position. If necessary, adjust paraffin block's alignment to the knife edge with the three leveling screws on the sides and bottom of the chuck.

5. Replace microtome knife holder on the microtome. Insert into the holder either a microtome knife or a razor blade in the razor blade holder; tighten the clamp screws. Have the knife set so that sectioning will begin at one end of the knife or razor blade; thus as the edge becomes dull, the position of the knife can be changed until the entire length is used.

6. Adjust height and tilt or angle of the knife (see Section C following).

7. Unlock the brake of the microtome and with the right hand holding the rotary handle; rotate handle clockwise until the face of the tissue block is at the level of knife edge. Simultaneously with the left hand, slide the microtome knife holder towards the face of the block until it almost touches the face of the tissue block; tighten the knife holder to the microtome. Slowly completely rotate the handle to see if the block just clears the knife edge. Make final adjustments with the crank handle or tissue feed carrier knob on the side of the microtome. If block clears the knife edge, sectioning of the tissue can begin.

B. Sectioning

1. To section, turn rotary handle with right hand in a clockwise direction at a moderate rhythmical pace. In the left hand hold a fine-tip paint brush. Brushes are preferred to other apparatus, as there is less chance of nicking the knife.

2. As the sections are made, they adhere to one another at one edge to form a ribbon. Friction due to sectioning of the paraffin generates enough heat to soften paraffin sufficiently for the sections to adhere. To get the ribbon started hold the first few sections down with a brush, then place the brush under the ribbon to support it off the knife edge. When the ribbon reaches about 15–20 cm (6–8 in.) long, stop sectioning, lock the rotary handle with the brake and with two brushes detach the ribbon from the knife edge.

3. Place the ribbon in a "ribbon box" dull side up, shiny side down for later attachment to glass slides (see Exercise 6). If ribbons are not immediately attached to the slides, store them in the "ribbon box" in a cool place so that they will not stick to the bottom of the box. A "ribbon box" may be any long low flat box with a cover, such as a tie box, stocking box or candy box. Line the bottom of the box with dark colored paper to make the sections of ribbon stand out more readily. When properly stored, the sections in the ribbons will keep their shape, but old sections will not stain as well as newly cut sections since oxidation by air causes tissue to deteriorate.

4. When you are finished sectioning, remove the knife! If you leave the microtome, remove the knife! Carelessness with the microtome knife can lead to serious accidents!

5. To aid sectioning the following suggestions should be kept in mind:

   a. The face of the paraffin tissue block must be hard, so keep face of the block cool with "cryokwik"[3] spray (Freon 12) or with an ice cube. Blot ice water from block and knife before sectioning.
   b. Cutting facets of the knife must be clean and free of tissue sections and paraffin. Brush knife edge and clean with a xylene-moistened rag. Especially do not allow paraffin to build up on back cutting facet, as this will prevent ribboning as well as cause splitting of the ribbon.
   c. Surrounding air must not be charged with static electricity, as this causes sections and ribbons to stick to the microtome. In dry weather this presents a problem, since friction of the knife as it cuts the paraffin block forms static electricity on the surface. The static electricity can be discharged with moisture (boil water to increase humidity of the environment) or with an anti-static irradiator or anti-static spray. The anti-static irradiator, overhanging the knife, emits alpha particles into the air to ionize the air and discharge the paraffin block's surface.

C. Adjustment of Height and Tilt or Angle of the Knife

1. **Height of the Knife.**—The proper height of the knife has the knife edge several millimeters beyond the jaws of the knife holder clamps. This brings the knife edge into proper relationship with the vertical movement of the tissue feed carrier mechanism. If the knife edge is too high, it will strike the lower edge of the paraffin block. Change of the knife height is not usual, unless knives of different widths are used. The height of the knife is adjusted with height adjusting screws at the bottom of the microtome knife holder before tightening the clamps of the knife holder.

2. **Tilt or Angle of the Knife.**—The proper tilt or angle of the knife depends on sharpness of the knife and hardness of the specimen of tissue. There is no easy way to determine the optimum tilt of the knife. It is found by "trial and error," even though the microtome knife holders may have angle markings on their sides.

[3]International Equipment Co., Needham Heights, Mass.

The optimum angle setting is 20° off the vertical for most sectioning, which leads to minimal compression of the tissue sections and formation of a ribbon. The angle is formed at the cutting edge between the center line of the knife and the vertical plane of the face of the tissue block (Fig. 5.8A). A properly tilted knife leaves a clearance angle of about 5°–10° between the back cutting facet of the knife and the face of the block, and just the knife edge strikes the face of the tissue block (Fig. 5.8A). If the knife is set in a too vertical position, the shoulder of the back cutting facet strikes the block before the knife edge strikes (Fig. 5.8B). This compresses the face of the tissue block as the block slides behind the knife and no sections result for several turns of the rotary handle. Then the block expands and a thick section is made. If the knife is set at too great an angle, the knife chisels or gouges through the tissue. The cut tissue sections have transverse folds and are destroyed (Fig. 5.8C).

After the clamps of the knife holder are tightened to hold the knife, the angle adjustment is made with the "wing" screws on the side of the microtome knife holder. Do not over-tighten any of the screws of the microtome. It is only necessary to have them finger tight.

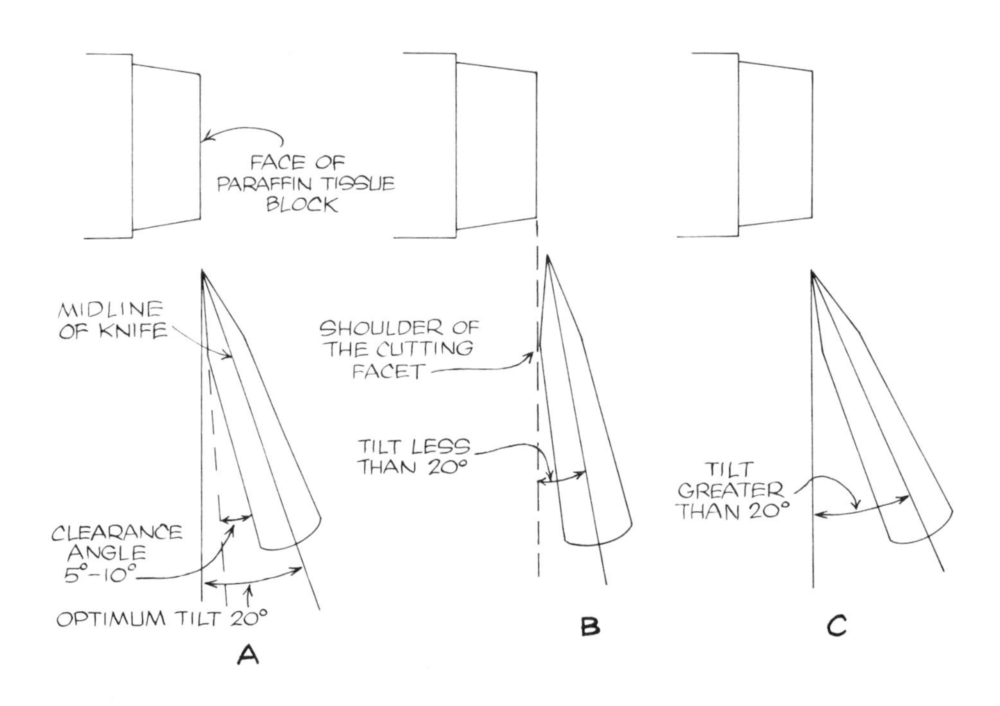

FIG. 5.8. TILT OF MICROTOME KNIFE.

A—Correct tilt: knife edge strikes the tissue block first, shoulder of cutting facet clears face of the block.
B—Insufficient tilt: knife too vertical, shoulder of cutting facet strikes the tissue block first instead of knife edge.
C—Excessive tilt: knife too horizontal, knife edge strikes the tissue block first but chisels through rather than cuts through the block.

IV. DIFFICULTIES COMMONLY ENCOUNTERED IN SECTIONING PARAFFIN EMBEDDED TISSUE AND SUGGESTED REMEDIES

A. Ribbon Fails to Form and Each Section Comes Off Separately

1. Upper and lower edges of paraffin block are not parallel to each other and to the knife edge: trim edges with razor blade and adjust knife edge parallel to face of the block.
2. Knife edge dirty due to buildup of paraffin on both cutting edges: clean edges by brushing and wiping with a xylene-moistened rag.
3. Knife angle too great, knife scrapes off the sections: tilt the angle of the knife less.
4. Knife edge is dull: resharpen edge or move to another part of the edge.
5. Paraffin too hard: use softer, lower melting point paraffin or soften paraffin with heat.
6. Temperature of the microtome knife too cold: warm the knife.
7. Sections are too thick: cut thinner sections.
8. Rate of sectioning too slow: section more rapidly.
9. To form a ribbon, unroll first section cut, hold it lightly against the knife with a brush. After the first few sections cut are held down, a ribbon will be formed.

B. Crooked and Uneven Ribbons

1. Upper and lower edges of paraffin block are not parallel to each other and to the knife edge (see Section A1).
2. Irregular or dull knife edge (see A4).
3. Uneven hardness of the paraffin or one side of the block is softer than the other: this is due to nonhomogeneous paraffin or heating one side of the block and not the other. Treat the first cause by reembedding tissue in homogeneous mixed paraffin. The second cause is treated by having uniform temperature of the environment during sectioning.

C. Sections Compressed, Wrinkled or Jammed Together

1. Knife is not sharp (see Section A4).
2. Knife angle insufficient: tilt the angle of the knife more.
3. Paraffin too soft: reembed tissue in harder paraffin or cool the block and knife.
4. Tissue not completely infiltrated and/or cleared: paraffin is mixed with clearing agents, makes the paraffin too soft. Return tissue to clearing agent, reinfiltrate and reembed.
5. Knife edge dirty (see A2).
6. Sections too thin or cut too rapidly: increase thickness setting or cut thin sections slowly and evenly.
7. Microtome screws are loose: tighten all screws only finger tight.

8. Another remedy for compressed sections is to soak the tissue block before sectioning in water or a mixture of glycerine and water (1:9) or 60% ethyl alcohol for 1 or 2 hr or overnight.

D. Sections Vary in Thickness or Are Skipped

1. Knife angle insufficient: tissue block is compressed, sections are skipped, then block expands and thick section is cut (see Section C2).
2. Knife angle too great (see A3).
3. Microtome screws are loose (see C7).
4. Microtome is worn due to lack of lubrication or is out of adjustment: have microtome serviced.
5. Paraffin blocks too large: may spring the knife edge; use smaller blocks.
6. Paraffin blocks with hard regions: may spring the knife edge (see C8).

E. Ribbon Splits or Has Vertical Scratches

1. Knife edge is nicked (see Section A4).
2. Knife angle too great (see A3).
3. Knife edge dirty or gritty (see A2).
4. Tissue too hard (see C8).
5. Dirt, crystals or other particles in paraffin or tissue. If paraffin is dirty, filter paraffin and reembed the tissue.

F. Tissue Sections Crumble or Fall Out of the Surrounding Embedded Paraffin

1. Incomplete dehydration of tissue: usually tissue can not be salvaged.
2. Incomplete clearing of tissue (see Section C4).
3. Tissue soft and mushy due to incomplete infiltration (see C4).
4. Tissue too hard: due to paraffin too hot for infiltration and embedding, or tissue too long in paraffin: tissue is "cooked" (see C8).
5. Tissue too hard due to length of time in clearing agent or to clearing agent used: reduce time in clearing agent; use another clearing agent or a mixture of clearing agent and clearing oil, such as cedarwood oil, or soak block to soften (see C8).
6. Tissue too hard for the paraffin: use a harder paraffin.

G. Sections Lift from the Knife and Cling to the Face of the Paraffin Block

1. Knife angle insufficient (see Section C2).
2. Paraffin too soft (see C3).
3. Knife edge is dull (see A4).
4. Knife edge dirty (see A2).

H. Sections Adhere to the Knife

1. Knife angle insufficient (see Section C2).
2. Knife edge is dull (see A4).
3. Knife edge dirty (see A2).

I. Knife Rings as It Passes over the Tissue and Tissue Sections Are Scratched

1. Knife angle either too great or insufficient: readjust to proper angle.
2. Tissue too hard (see Section C8).
3. Knife too thin: use a heavier knife.
4. Tissue too hard for paraffin technique: use another infiltration and embedding procedure.

J. Tissue Makes Scratching Noise During Sectioning

1. Tissue too hard (see Section C8).
2. Tissue too hard for paraffin technique (see I4).
3. Crystals of fixative or calcium deposits, etc., in the tissue: caused by improper removal during the processing of the tissue.
4. Dirt in the paraffin (see E5).

K. Undulations in Surface of the Section

1. Microtome screws are loose (see Section C7).
2. Knife angle too great (see A3).

L. Sections Curl, Fly, Stick to Parts of the Microtome or Other Nearby Objects due to Static Electricity from Friction of Sectioning. This Occurs During Conditions of Low Humidity

1. Increase room humidity by boiling water in an open shallow pan.
2. Use anti-static instrumentation to ionize the air.
3. Attach ground wire or chain to microtome clamp.
4. Make short ribbons.
5. Postpone sectioning until weather is more humid: section in the morning hours when humidity of the environment is higher.

QUESTIONS

Cover answers with a piece of paper. Answers appear at end of questions.

(1) In a rotary microtome, the tissue block, in relation to the cutting edge of the knife, is
- (a) In the horizontal plane at right angles to the knife
- (b) In the vertical plane at 30° angle to the knife
- (c) In a hortizontal plane at 180° to the knife
- (d) In a vertical plane at 360° to the knife

(2) In principle of operation a microtome is very similar to
- (a) An automatic tissue processor
- (b) A hone
- (c) A strop
- (d) A delicatessen meat slicer

(3) Which of the following is not an advantage of using razor blades instead of a microtome knife?
- (a) No sharpening required
- (b) Blades are more rigid than knife
- (c) Razor blades are cheaper
- (d) Cutting edge of razor does not enter beyond the ends of the knife holder

(4) Which of the following statements is correct?
- (a) Stropping precedes honing
- (b) Honing uses very fine abrasives
- (c) All honing must be done manually
- (d) Honing is done before stropping

(5) Neat's-foot oil is used
- (a) To retain a strop leather's pliability
- (b) To clean a strop
- (c) To oil a hone
- (d) As an abrasive for honing

(6) Routine sections are cut
- (a) Between 6 and 10 microns
- (b) Between 2 and 3 microns
- (c) Between 10 and 20 microns
- (d) Only at 5 microns

(7) A "ribbon" should be stopped when it reaches
- (a) Melting temperature
- (b) 15–20 cm in length
- (c) 12 in. in length
- (d) The table top

(8) "Cryokwik" spray (Freon 12) is used to
- (a) Lubricate the strop
- (b) Lubricate the hone
- (c) Keep the face of the block cool
- (d) Remove ice water from block

(9) The proper height of the knife is
- (a) Knife edge several millimeters beyond jaws of knife holder clamps
- (b) Knife edge several millimeters below jaws of knife holder clamps
- (c) Knife edge several inches beyond jaws of knife holder clamps
- (d) Knife edge several inches below jaws of knife holder clamps

(10) The best way to determine the proper tilt or angle of the blade is
- (a) Arbitrarily set at 20°
- (b) Have a clearance angle of 18°
- (c) Trial and error
- (d) Degree of curling of ribbon

Answers

(1) a
(2) d
(3) b
(4) d
(5) a
(6) a
(7) b
(8) c
(9) a
(10) c

6

Spreading the Sections and Attachment or Mounting of Sections to Glass Slides

The tissues must be attached to a solid surface for support as the tissues fall apart upon removal of the paraffin prior to staining. The tissues are attached to glass slides which give them support.

There are several techniques used to flatten or spread the paraffin embedded tissue sections and to attach them to glass slides. For routine histological technique, 3 × 1 in. (75 × 25 mm, 1 mm thick) slides are used and should be perfectly clean. Precleaned slides ready for use save time; otherwise clean slides by dipping them into a 1% acid-alcohol solution (1 ml of concentrated HCl to 99 ml of 95% alcohol), and wipe dry with a lint-free cloth.

The most important thing that should be done to the slides is to inscribe at one edge, with a scriber, the code number of the tissues that will be attached to these slides. This should be performed before the tissue sections are affixed.

To center the sections on the slide, a model of the slide should be drawn out on an index card (see Fig. 6.1). The slide is placed over the model and the sections can then be centered.

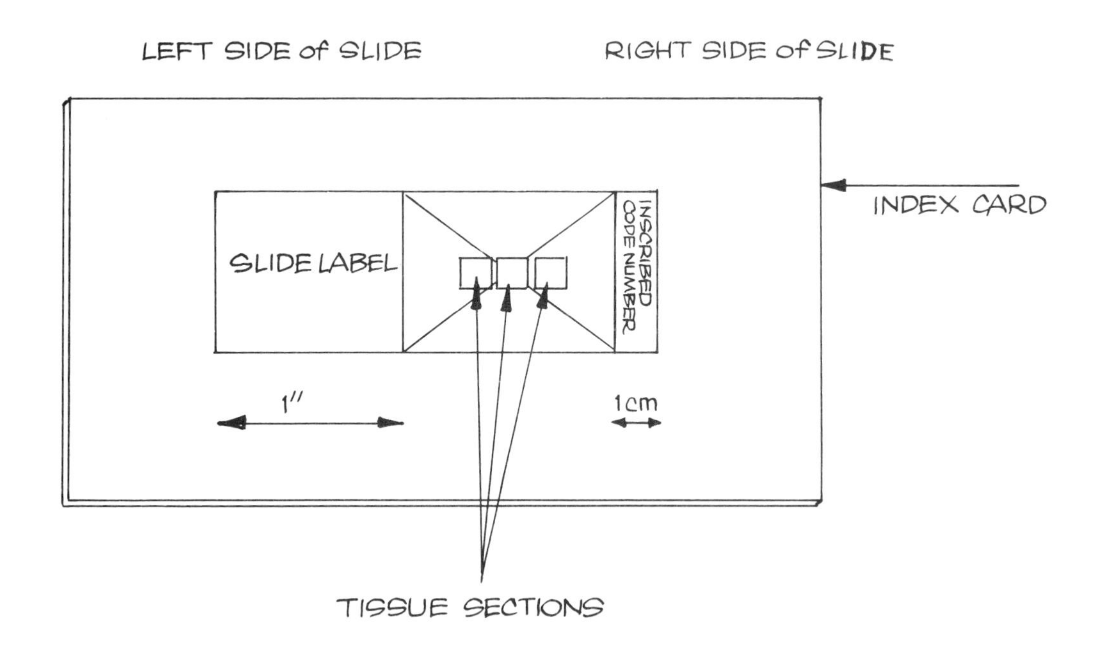

FIG. 6.1. CENTERING SECTIONS ON THE SLIDE

I. SPREADING THE SECTIONS

During sectioning, the sections become compressed and wrinkled and have to be flattened before they are firmly affixed to the slides. The spreading of the sections takes place by floating the ribbons of tissue sections on warm distilled water at a temperature about 5°–10°C below the melting point of the paraffin. A water temperature of 45°–50°C will soften the paraffin to spread the sections and remove the wrinkles. The temperature should not melt the paraffin as this can cause shrinkage, tearing and displacement of tissue components. If temperature is not warm enough, tissue will not flatten properly.

II. ATTACHMENT OR MOUNTING OF SECTIONS TO GLASS SLIDES

A. Warming Plate Method

This is the classical manner of attaching tissue sections to the slide. The adhesive generally used for attachment of the tissue sections is Mayer's egg albumin, which can be easily made or purchased already prepared. This solution should be stored in the refrigerator to prevent bacterial and mold growth and should keep for several months. The Mayer's egg albumin consists of:

Egg white	50 ml
Glycerin	50 ml
Sodium salicylate or thymol as a preservative	1 g

For current use, dispense small amounts (1 ml) of the albumin from the stock bottle into small containers with glass rod applicators. This avoids contamination of the stock. The small containers should be well washed before refilling with fresh albumin. There are other adhesives that might also be used, such as gelatin, gum, blood serum, "subbing" solution. Some types of tissue sections can be affixed to slides without adhesive, but there is a risk that the sections will fall off during the staining procedures. Some histologists suggest that albumin does not act as an adhesive, but as a surface tension depressant which aids closer attraction of sections to the slide.

Using clean, pre-coded slides, a minute drop of albumin is applied to the surface of the slide on which the code number is inscribed. Then with a fingertip spread a thin, even coat of the adhesive on this surface. Wipe excess albumin off with another finger. Excess albumin will cause the sections to come off the slide, will pick up stain, and will make the slide appear unsightly, preventing a clear view of the tissue on examination.

Paraffin ribbons collected in the ribbon box are cut into smaller units (tissue sections of 3–5 units long). Several drops of distilled water are placed over the albumin. The tissue sections are picked up with a brush and floated on the water, with their shiny side against the slide. The sections are centered on the slide using the index card (see Fig. 6.1). The slide is then placed on the warming plate to spread the sections. As the paraffin softens, the ribbon of sections may be gently pulled with clean needles to stretch it. The slides are kept on the warming plate until sections are flat and set firmly against the slide. The excess water can be drained off. If air bubbles are trapped under the sections, they can be dislodged by brushing the bubbles gently toward the edges of the section before the section sets firmly against the slide.

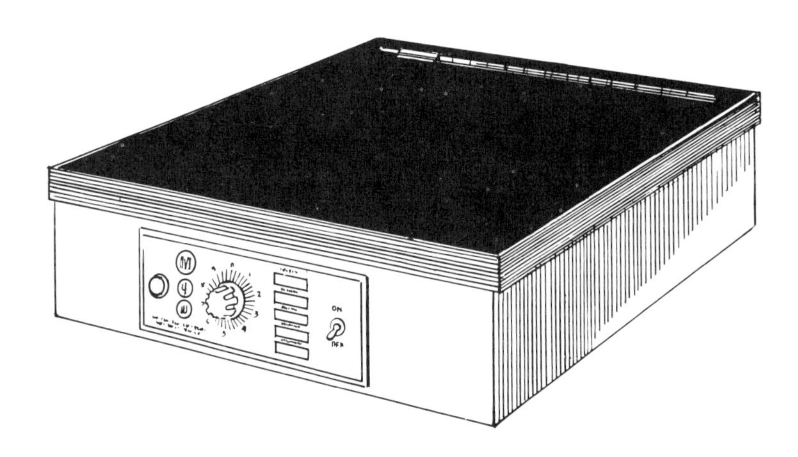

FIG. 6.2. WARMING PLATE FOR SLIDES

B. Flotation Method in a Water Bath

The tissue flotation water bath is electrically heated and temperature controlled. The baths have a black painted interior so tissue sections can be easily seen. The adhesive is usually gelatin, which has been shown to be a better adhesive than albumin for this method and does not pick up the stain, but Mayer's albumin may be used. A 1/4 teaspoon of U.S.P. gelatin is sprinkled and stirred into the warm water bath of distilled water before flotation of the paraffin ribbons.

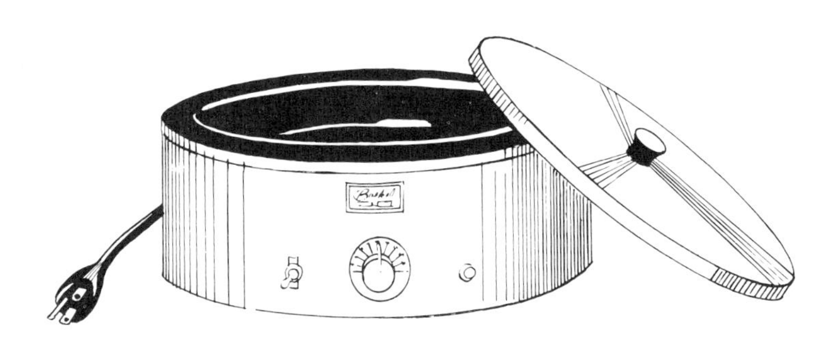

FIG. 6.3. TISSUE FLOTATION BATH

To float a ribbon, pick up one end of the ribbon (shiny surface down) with a brush and let the other end contact the water slowly so wetting occurs throughout the length of the ribbon. If ribbon is dropped flat on the surface, folds may develop and air bubbles will be trapped under the sections. The air bubbles may be removed, but with difficulty. Another source of air bubbles is from dissolved gases in the water which escape on heating. Boil water to drive off the gases; cool before using.

After the paraffin has softened, the ribbon can be gently stretched. Then with a rapid, deliberate stroke, using a hot, clean scalpel, the ribbon can be cut into small units (3–5 sections long). If there are still hard to remove wrinkles, add a few drops of 60–80% alcohol to the water close to the ribbons.

Dip the clean, pre-coded slide under the tissue sections, draw up sections to slide with a brush, withdraw slide from the water. Center the sections, drain excess water and place on warming plate to firmly set the tissue sections against the slides.

The surface of the water should be frequently cleaned by skimming with a sheet of paper. Fresh water and adhesive should be used daily.

When tissue sections are firmly attached to the slides, permit slides to dry overnight or in a hot air dryer for a few hours before staining them.

Properly attached sections will have a smooth, clear appearance. If they are opaque, air is trapped between sections and glass and these sections will have a greater tendency to be lost during the staining procedure. The causes of loose sections are poorly cleaned slides, drying slides at too low a temperature, not drying the slides long enough, or air bubbles trapped under the sections. If sections have a tendency to float off during the staining, they can be coated with a thin film of 0.5–1.0% nitrocellulose or celloidin solution.

QUESTIONS

Cover answers with a piece of paper. Answers appear at end of questions.

(1) Spreading of sections is necessary because
- (a) Cells become separated
- (b) Tissue is restored to original shape
- (c) Sections are compressed
- (d) Staining is enhanced

(2) Spreading of sections can be done by floating on distilled water at
- (a) 45°–50°C
- (b) 20°C below melting point of paraffin
- (c) 25°C above melting point of paraffin
- (d) 25°–30°C

(3) The two main ingredients of Mayer's adhesive are
- (a) H_2O and thymol
- (b) Thymol and egg white
- (c) Egg white and glycerin
- (d) Egg white and sodium salicylate

(4) Ribbons can easily be cut into small units (3–5 sections long) by use of
- (a) Scissors
- (b) Fingers
- (c) Hot water bath
- (d) A hot clean scalpel

(5) If sections on a slide appear opaque, it is probably due to
- (a) Too much adhesive
- (b) Too little adhesive
- (c) Sections being too thin
- (d) Air trapped between sections and slide

(6) If sections tend to float off during staining, sections can be coated with
- (a) Thymol and sodium salicylate
- (b) 0.5–1.0% Nitrocellulose or celloidin solution
- (c) Egg white
- (d) Glycerin

(7) The use of an electrically heated tissue flotation bath has advantages, two of which would be
- (a) No water required
- (b) Slides do not get excessively wet
- (c) Better temperature control
- (d) Sections more easily seen

(8) Slides, if not precleaned, should be dipped in
- (a) Distilled H_2O
- (b) Absolute methyl alcohol
- (c) Albumin
- (d) 1% Acid-alcohol solution

(9) It is ABSOLUTELY imperative that
- (a) The code number be inscribed on the slide for the tissue to be attached to that slide
- (b) Mayer's adhesive be used
- (c) Sections be perfectly centered
- (d) 3 mm thick slides be used

(10) Air bubbles entrapped under sections can be dislodged by
- (a) Refloating section
- (b) Poking with a teasing needle
- (c) Heating slide to 60°C
- (d) Brushing toward edges of the section

Answers

(1) c
(2) a
(3) c
(4) d
(5) d
(6) b
(7) c and d
(8) d
(9) a
(10) d

7

General Staining Procedure for Paraffin Infiltrated and Embedded Tissue

The purpose of staining the tissue sections is to impart color to the tissue components which do not retain enough color after processing. The use of dyes or stains makes the tissue components visible by bringing out differences in refractive indexes.

I. NUCLEAR STAINS

Hematoxylin is an example of a nuclear stain. It is a natural stain obtained by extraction from the bark of the logwood tree of Central and South America. Hematoxylin was one of the first and is still one of the more widely used stains in histology. Other examples of natural stains are carmine and brazilin. Hematoxylin is not the active form of the stain; it must be oxidized to hematein. This process is known as "ripening" the stain. Oxidation or ripening can take place slowly by air oxidation for a period of 6–8 weeks. The oxidation can also rapidly take place by addition of an oxidizing agent, such as sodium iodate, mercuric oxide, potassium permanganate or hydrogen peroxide.

If hematoxylin is oxidized without an oxidizing agent, the pH of the solvent and type of solvent determine the rate of conversion of hematoxylin to hematein (Table 7.1).

If you want to preserve a hematoxylin solution for a long time, it should be acidified and made up in alcohol to which glycerin has been added. Alcoholic solutions of hematoxylin last five times longer than aqueous solutions.

Solutions of hematoxylin are used rather than hematein, since hematein rapidly loses its staining efficiency, due to its continued oxidation to products which precipitate out of solution. It is better to have hematoxylin present to replenish the hematein lost by its further oxidation.

The color changes that take place in the hematoxylin solution indicate its staining efficiency. The changes are from white to lilac, bright purple, deep purple, red, orange-red, orange-brown and brown. When the hematoxylin solution is brown, all the hematoxylin has been converted to hematein and the solution is no longer useful. The hematoxylin solution stains best when its color is in the purple range, as the hematoxylin is partially oxidized to hematein, with sufficient hematoxylin present to replace oxidized hematein.

TABLE 7.1
EFFECT OF pH AND TYPE OF SOLVENT ON RATE OF CONVERSION OF HEMATOXYLIN TO HEMATEIN

pH and Type of Solvent	Rate of Oxidation of Hematoxylin to Hematein in Aqueous Neutral Solvent as Compared to Other Solvents
1. Aqueous solvent	
a. Neutral	Hematein forms in a few hours
b. Acid	Hematein forms more slowly
c. Alkaline	Hematein forms more rapidly
2. Alcoholic solvent	Hematein forms more slowly
3. Alcoholic solvent plus glycerin	Hematein forms even more slowly

To test the efficiency of a hematoxylin solution, add several drops of the hematoxylin solution to tap water. Note the color change of the tap water: if it turns blue-purple immediately, it contains the proper ratio of hematoxylin to hematein. If it changes slowly or stays red or brown, it has too much hematein or the hematein has been further oxidized and the solution should be discarded.

Hematein by itself has little affinity or attraction for the tissue, unless it is attached to a substance which is called a mordant. A mordant acts to increase the attachment of the stain (hematein) to the tissue. The mordant also will determine if the hematoxylin stains the nuclei blue or black. Mordants that contain iron will cause the hematoxylin to stain the nuclei black; mordants containing other metallic ions such as aluminum or potassium or ammonium will stain the nuclei blue to purple. If a mordant contains sulfate ions, the age of the mordanted hematoxylin determines its staining efficiency, as the sulfate ion in solution forms sulfuric acid and increases the oxidation of hematoxylin to hematein.

Classification of hematoxylin solutions depends on mordanted hematoxylin being mixed or not mixed. If the mordant and the hematoxylin are mixed, this is called a "direct" hematoxylin stain. If the mordant precedes the hematoxylin, this is called "mordant" hematoxylin stain.

A. "Direct" Hematoxylin Stains

1. Alum Mordant Hematoxylins.—These stains consist of hematoxylin and salts of aluminum, which is the mordant. Mordants that contain aluminum salts are called alums.

a. Delafield's Hematoxylin.—

Hematoxylin (C.I. 75290)[1]	4.0 g
Ethyl alcohol, 95% or absolute	25.0 ml

[1]C.I. Number—The color index number is a specific number used to reference a given stain that has been certified by the Biological Stain Commission.

Ammonium alum ($NH_4Al(SO_4)_2 \cdot 12H_2O$) saturated aqueous solution (15 g/100 ml)—mordant	400.0 ml

Dissolve the hematoxylin in the alcohol. Add this to the mordant. Leave exposed to light and air to ripen in a clear, cotton-plugged flask for 3–5 days. Filter.

To the filtrate add:

Glycerin	100.0 ml
Ethyl or methyl alcohol, 95% or absolute	100.0 ml

Allow to ripen for about 6 weeks and test solution for staining efficiency as described above every few weeks. When the correct color of the stain is obtained, store in a tightly capped bottle. In this way the ripened solution can be kept for years.

b. **Ehrlich's Hematoxylin.—**

Hematoxylin (C.I. 75290)	2.0 g
Ammonium alum	3.0 g
Ethyl or methyl alcohol, 95% or absolute	100.0 ml
Glycerin	100.0 ml
Distilled water	100.0 ml
Glacial acetic acid	10.0 ml

Dissolve the hematoxylin in the alcohol, then add the other reagents. The solution ripens by exposure to light and air in a cotton-plugged flask for two weeks or longer; stir frequently. Solution may be ripened immediately by addition of 0.3–0.4 g of sodium iodate ($NaIO_3$), an oxidizing agent. This partially oxidizes some of the hematoxylin to hematein. After ripening, store in a tightly capped bottle. Solution keeps for years.

c. **Harris' Hematoxylin.—**

Hematoxylin (C.I. 75290)	1.0 g
Ethyl alcohol absolute	10.0 ml
Potassium alum ($KAl(SO_4)_2 \cdot 12H_2O$) or ammonium alum	20.0 g
Distilled water	200.0 ml
Mercuric oxide (HgO)	0.5 g
Glacial acetic acid	8.0 ml

Dissolve the hematoxylin in the alcohol. Dissolve the alum in hot boiling water in a large flask. Add the hematoxylin solution to the alum; boil for about 0.5 min. Place the flask in a basin of cold water, cool rapidly and add the mercuric oxide slowly, as the reaction can be explosive. Keep the flask in the cold water until solution turns dark purple. When cool add the glacial acetic acid to accentuate the stain, to keep away metallic luster and to prevent staining of the cytoplasmic components. Allow stain to ripen for 1 week; test for efficiency before use. Solution keeps well for 1 to 2 months. Filter before using.

d. **Mayer's Hematoxylin.**—

Hematoxylin (C.I. 75290)	1.0 g
Sodium iodate	0.2 g
Potassium or ammonium alum	50.0 g
Citric acid	1.0 g
Chloral hydrate	50.0 g
Distilled water	1000.0 ml

Dissolve the hematoxylin in water using gentle heat. Then add the reagents in the order listed above. The final color of the solution is red-violet. Allow to ripen 6–8 weeks, although stain may be used within 1–2 weeks. The citric acid brightens the staining action and chloral hydrate acts as a preservative. The stain keeps well for years.

2. **Iron Mordant Hematoxylins.**—

a. **Weigert's Iron Hematoxylin.**—(Lilly's Modification, 1968)

Solution A—Mordant

Ferric chloride ($FeCl_3 \cdot 6H_2O$)	2.5 g
Ferrous sulfate ($FeSO_4 \cdot 7H_2O$)	4.5 g
Hydrochloric acid, concentrated	2.0 ml
Distilled water	298.0 ml

Solution B—Stain

Hematoxylin (C.I. 75290)	1.0 g
Ethyl alcohol, 95%	100.0 ml

Solutions A and B are stored separately. If hematoxylin solution is fresh, mix equal parts of Solutions A and B just before use; solution turns black at once. If hematoxylin is not fresh, use less of Solution A. Best results are obtained when solutions are mixed fresh each time, but mixture is usable for about 2–3 weeks after mixing or until solution turns brown.

b. **Groat's Variation of Weigert's Iron Hematoxylin.**—

Distilled water	50.0 ml
Sulfuric acid, concentrated	0.8 ml
Ferric alum ($FeNH_4(SO_4)_2 \cdot 12H_2O$)	1.0 g
Ethyl alcohol, 95%	50.0 ml
Hematoxylin (C.I. 75290)	0.5 g

This variation is a single solution. Mix reagents in the order given at room temperature and filter.

B. "Mordant" Hematoxylin Stains—Iron Mordants

1. Heidenhain's Iron Hematoxylin.—

Solution A—Mordant

Ferric ammonium alum	4.0 g
Distilled water	100.0 ml

Solution B—Stain

Hematoxylin (C.I. 75290)	0.5 g
Ethyl alcohol, 95%	10.0 ml
Distilled water	90.0 ml

Dissolve the hematoxylin in alcohol, add the water, allow to ripen 4–5 weeks, store in tightly capped bottle. Solutions A and B are not mixed. If they are, stain rapidly deteriorates. In a staining procedure, Solution A, the mordant, precedes Solution B, the stain.

2. Mallory's Iron Chloride Hematoxylin.—

Solution A—Mordant

Ferric chloride	5.0 g
Distilled water	100.0 ml

Solution B—Stain

Hematoxylin (C.I. 75290)	0.5 g
Distilled water	100.0 ml

Prepare fresh each time the stain is used. Never mix the two solutions together. Solution A precedes Solution B in a staining procedure.

II. PLASMA STAINS

These stains will stain the cytoplasm of the cell and other tissue components. These stains are mainly synthetic stains, although a few are natural, for example, indigo carmine. The plasma stains are counterstains to the nuclear stains.

Eosin is the most common and widely used. It is used either as an aqueous or alcoholic solution at a concentration of 0.5–1%. Eosin stains rapidly and brilliantly depending on the fixative used and pH of the eosin solution. Mercuric salts of Zenker's fixative attach to the negatively charged groups of the tissue components. This increases the attraction of the negatively charged eosin stain to the positively charged groups of the tissue components. For formalin fixed tissue, it is suggested to place the tissue sections in a saturated solution of mercuric chloride for 2–3 min prior to counterstaining with eosin. This should increase the tissue attraction for the eosin as per reasons stated before.

Eosin will stain best if its solution is slightly acidic. Under these conditions, the tissue components are positively charged and will attract negatively charged eosin. The optimum pH for eosin solution is 5.4–5.6. The acidified counterstain gives good results for about 2 weeks, then pH will rise and the stain is no longer useful.

Eosin is rapidly removed during dehydration in 95% and absolute alcohol. Thus tissue sections should be slightly overstained and rapidly dehydrated through the alcohols to the clearing agent.

A. Alcoholic Eosin

Eosin Y (yellowish) (C.I. 45380) or Eosin B (bluish) (C.I. 45400)	0.5 g
Ethyl alcohol, 95%	100.0 ml
Glacial acetic acid	0.5 ml

Eosin may be made up in other alcohol percentages. These then should be in the correct place in the dehydration series of a staining procedure.

B. Aqueous Eosin

Eosin Y or Eosin B	1.0 g
Distilled water	100.0 ml
Glacial acetic acid	0.5 ml

C. Eosin—Orange G

Eosin Y, 1% solution in 95% ethyl alcohol	10.0 ml
Orange G (C.I. 46230) saturated 95% ethyl alcohol solution (0.5 g/100 ml)	5.0 ml
Ethyl alcohol, 95%	45.0 ml

D. Other Counterstains

Name of Stain	Color Index (C.I.)	Color	Type of Solution
1. Acid fuchsin	42685	Magenta	5% Aqueous solution
2. Aniline blue, WS	42780	Blue	0.5–2.0% Aqueous solution
3. Biebrich scarlet	26905	Red	1% Aqueous solution
4. Congo red	22120	Red	0.5% Aqueous solution

Name of Stain	Color Index (C.I.)	Color	Type of Solution
5. Fast green, FCF	42053	Green	0.2–0.3% Aqueous solution or 95% ethyl alcohol
6. Light green, SF (yellowish)	42095	Yellow	0.2–0.3% Aqueous solution or 95% ethyl alcohol
7. Orange G	16230	Orange	0.5% in 95% Ethyl alcohol
8. Phloxine B	45410	Red-purple	0.5% Aqueous solution plus 0.2 ml glacial acetic acid

III. EQUIPMENT NEEDED FOR MANUAL STAINING PROCEDURES

For staining of tissue sections attached to glass slides, Coplin jars with plastic holders to accommodate the slides or staining dishes with slide racks are generally used. For staining sections mounted on cover slips, Columbia jars are used. Covers for the containers should be tight fitting to prevent evaporation of reagents, absorption of moisture, and contamination from other sources. Containers should be clearly labeled and the solutions should be renewed when they appear to be contaminated. The containers should be about 2/3 to 3/4 full and not filled to the top.

When loading the plastic holder with the slides for staining in the Coplin jars, make sure code number is up and tissue is facing you. Once the paraffin is removed, it is difficult to tell on which side the tissue sections are placed.

IV. BASIC PROCEDURES FOR STAINING PARAFFIN INFILTRATED AND EMBEDDED SECTIONS

A. Deparaffinization

Since the stain, which is usually water soluble, can not permeate the paraffin, the paraffin is removed with a highly organic solvent such as xylene. Usually two changes of xylene are sufficient. The second xylene assures removal of any paraffin remaining in the tissue sections. The time in xylene is about 2 min each, but check to see if all the paraffin has been dissolved; if not keep tissue sections in xylene longer. Xylene is the solvent usually used for deparaffinization and clearing (see Section G), but toluene, benzene or other similar organic chemicals could be employed.

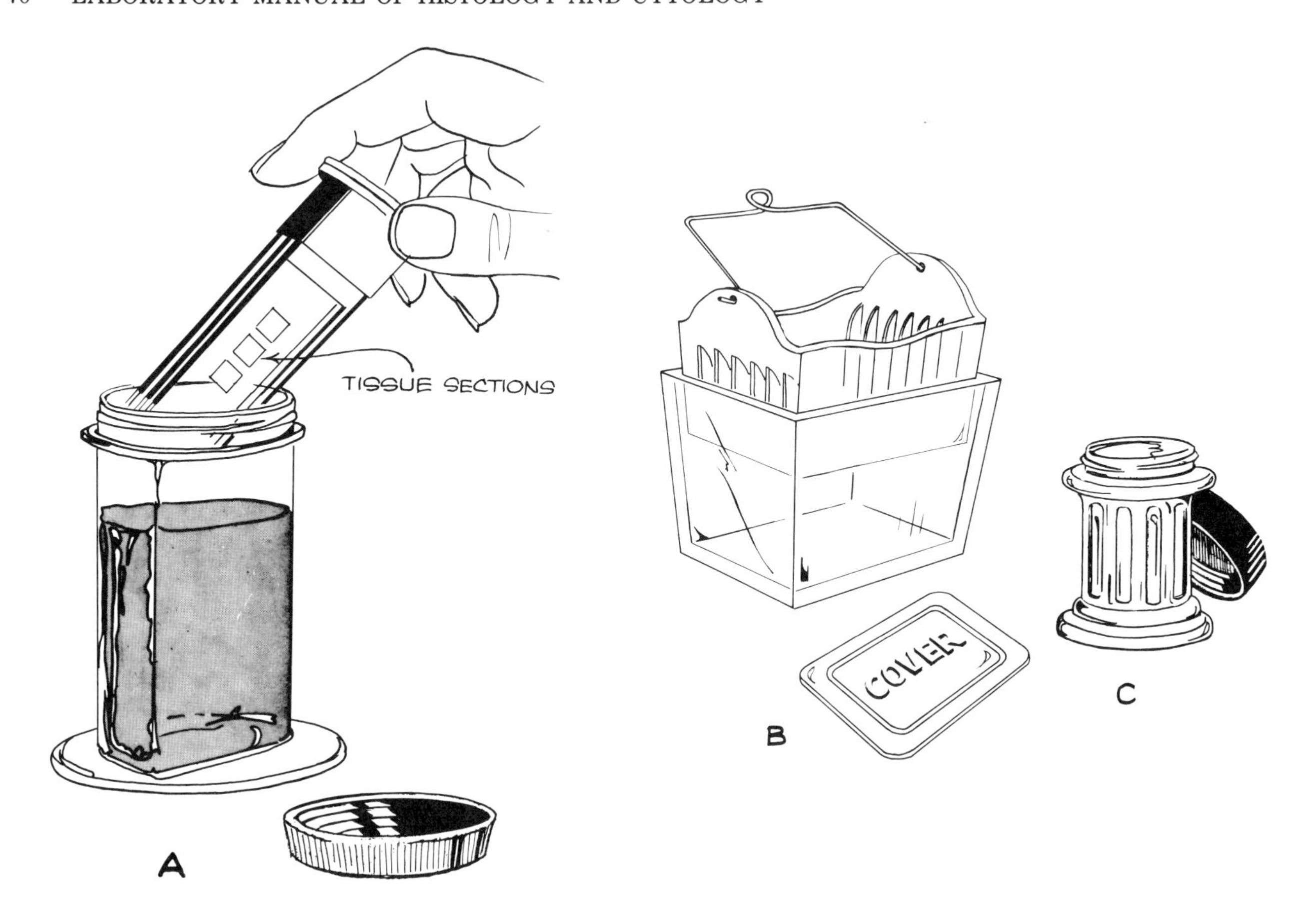

FIG. 7.1. CONTAINERS AND HOLDERS FOR MANUAL STAINING.
A—Coplin jar and plastic slide holder.
B—Staining dish and rack.
C—Columbia jar.

Before transferring slides from one solution to the next in the staining procedure, drain and blot edges of slides to minimize carry-over and contamination. Once the paraffin is removed, never allow tissue sections to dry.

B. Absolute Alcohol

Xylene is removed from the tissue sections with absolute alcohol since xylene is not miscible with lower concentrations of alcohol and water. One or two changes of absolute alcohol for about 1 min each is adequate.

C. Hydration with a Series of Gradually Decreasing Concentrations of Alcohols

To prevent rapid movement of fluids into and out of the tissue sections and consequent loss of the sections from the glass slides, the sections should be "run" gradually through a series of decreasing concentrations of alcohols (95, 80, 70% and, for delicate tissues, even 50 and 30%) to water. The tissue sections should be left in each alcohol concentration for about 1 min.

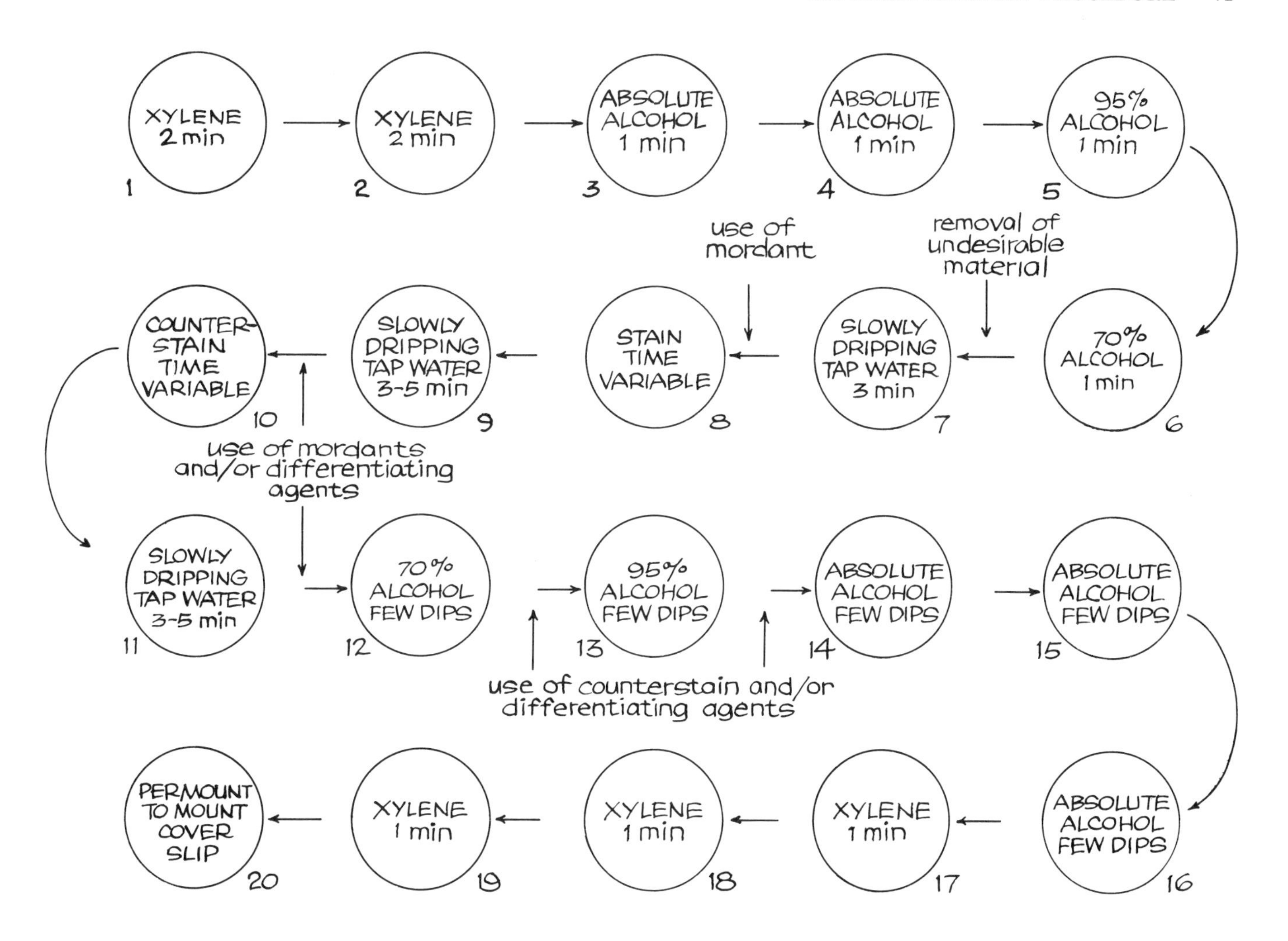

FIG. 7.2. GENERAL ARRANGEMENT OF COPLIN JARS FOR A MANUAL STAINING PROCEDURE

During dehydration undesirable pigments and other materials are removed. For example, with tissue fixed in mercuric chloride-containing fixatives such as Zenker's fixative or Technicon's FU-48, the mercury precipitates as black clumps and needles which must be removed before staining. The black clumps or needles, called mercuric pigment, are made soluble by placing the tissue sections into an alcoholic-iodine solution (Lugol's solution) for about 3 min. The tissues are then washed in distilled water to remove the excess iodine for about 3 min. The iodine is decolorized or "bleached" with 5% sodium thiosulfate for about 2 min and the thiosulfate is washed out in slowly dripping tap water for about 3 min. Use slowly dripping water to prevent tissue sections from being loosened from the slides.

The alcohol usually used for hydration and dehydration (see Section F) is ethyl alcohol, but isopropyl alcohol or tertiary butyl alcohol may be employed.

D. Washing

This step is carried out if tissue sections have not already been brought to a comparable solvent. During this step, the tissue sections are brought to the solvent of the stain, usually water, or the concentration of alcohol in which the stain is dissolved. If water is the solvent, the washing is accomplished in slowly dripping tap water for about 3 min.

E. Staining and Counterstaining

To obtain properly stained tissue sections, the order of applying the stains and counterstains and the use of mordants, differentiating, or decolorizing agents must be in the prescribed manner with water washes between steps to prevent carry-over of one solution to the next. The time for these steps is variable depending on desired staining intensity.

The use of the counterstains and differentiating agents allows differences among the tissue components to be demonstrated. The mordant acts to hold the stain strongly to the tissue in order to prevent the stain from being lost from the tissue sections when they are placed in water or low concentrations of alcohol, as in the dehydration sequence (see Section F).

F. Dehydration with a Series of Gradually Increasing Concentrations of Alcohols

and

G. Absolute Alcohol

In routine histological technique, stained tissue sections are mounted in a mounting medium soluble in an organic solvent, such as xylene. The tissue sections must be dehydrated in a series of gradually increasing concentrations of alcohols (70, 95%, 2 or 3 changes of 100%) before passing into xylene. Unless tissue sections are completely dehydrated, tissue sections will not become transparent when placed in xylene. The mounting medium will become cloudy or opaque, as water does not mix with the mounting medium and stain may fade.

The time in each alcohol concentration is very short, just a few dips, as many stains are soluble in low alcohol concentrations and can be extracted from the tissue sections during this sequence of steps. It is also possible to include in this series alcoholic solutions of counterstains and differentiating fluids.

H. Clearing

The tissue sections are passed through two or three changes of an organic fluid such as xylene or toluene for about 1 min each to clear or to make the tissue sections transparent and assure the removal of the alcohol. The xylene also serves as the solvent for the mounting medium.

It is preferable to set up a separate series of containers for dehydration and clearing rather than to use the same ones for deparaffinization and hydration. For example, if the series of containers were not separate, the paraffin in the xylene used for deparaffinization would interfere with mounting. Or the absolute alcohol, used in dehydration series, could become diluted due to carry-over from tissue sections in lower concentrations of alcohol. This alcohol would not be suitable to use after deparaffinization since it would be contaminated with water and would not remove xylene from the tissue sections.

I. Mounting the Cover Slip

To make the slide a permanent preparation, the mounting medium is applied to the cover slip and the cover slip is placed over the tissue sections (see Exercise 8).

TABLE 7.2
SUMMARY OF GENERAL PROCEDURE FOR STAINING

A. Deparaffinization
1. Xylene—2 min or until all visible paraffin is removed
2. Xylene—2 min

B. Absolute Alcohol
3. Absolute alcohol—1 min
4. Absolute alcohol—1 min

C. Hydration
5. 95% Alcohol—1 min
6. 70% Alcohol—1 min

Undesirable pigment and other material should be removed during these steps.

D. Washing
7. Slowly dripping tap water—3 min

E. Staining and Counterstaining
8. Stain
9. Slowly dripping tap water—3–5 min
10. Counterstain
11. Slowly dripping tap water—3–5 min

The time for stain and counterstain is variable, depending on the desired stain intensity. Mordants and differentiating agents are used during these steps.

F. Dehydration
12. 70% Alcohol—few dips
13. 95% Alcohol—few dips

Also counterstaining and differentiation may occur during these steps.

G. Absolute Alcohol
14. Absolute alcohol—few dips
15. Absolute alcohol—few dips
16. Absolute alcohol—few dips

H. Clearing
17. Xylene—1 min
18. Xylene—1 min
19. Xylene—1 min

I. Mounting the Cover Slip

The procedure outlined in Table 7.2 and Fig. 7.2 is general and these steps may have to be modified for specific staining.

In many staining procedures where the steps at the beginning and end are much the same, there is a shortening of the description, so that the statements “deparaffinize or decerate and hydrate or run slides to water” are used to refer to Steps A through D of the general procedure. “Dehydrate or run slides up, clear and mount in the usual manner” refers to Steps F through I. If there are any special steps in a staining procedure, those parts of the procedure are described in detail as well as the application of the specific stains, counterstains and other specific reagents, such as mordants, differentiating agents, etc.

Any staining procedure should list as closely as possible the following steps.

1. Tissue that can be stained by the method, if the method is specific for that tissue.
2. Types of fixatives that can be used.
3. Type of sections and thickness of sections used. Since paraffin method is routine in histotechnique, the staining procedures to be discussed are related to paraffin infiltrated and embedded tissue sections and section thickness between 5 and 10 microns.
4. Solutions and reagents used in the staining procedure.
5. Staining procedure itself:
   a. Preparation for staining—deparaffinization and hydration
   b. Staining and counterstaining
   c. Preparation for making stained tissue sections permanent—dehydration, clearing, mounting
6. Results—color of the various tissue components.

V. ROUTINE HEMATOXYLIN AND EOSIN STAINING PROCEDURES

These staining procedures are used to distinguish between nucleus and cytoplasm of a cell and other tissue components to show general relationships of the tissues.

A. Progressive Methods

In these methods intensity of staining of the tissue sections increases with time.

1. Long Manual Procedure.—

a. **Fixative.**—Any general or routine fixative.

b. **Solutions.**—

(1) *Hematoxylin.*—(See preceding for formulae of various types of hematoxylin solutions.)

(2) *Eosin.*—(See preceding for formulae of various types of eosin solutions.)

c. **Procedure.**—

(1) Deparaffinize or decerate and hydrate or run slides down to water. Remove mercuric pigments if present.

(2)	Hematoxylin stain Time depends on strength of stain. After each minute or less, check under the microscope for correct intensity of stain. Rinse in slowly dripping tap water to stop staining action; blot bottom of slide before checking under the microscope. If stain is not dark enough, return slides to the stain.	1–5 min
(3)	When correct intensity of stain is obtained, wash in slowly dripping tap water	3–5 min
(4)	Blue the hematoxylin (except iron hematoxylin) in an alkaline solution such as ammonia water (a few drops of ammonia in tap water) or solution of sodium or lithium carbonate	3–5 min
(5)	Wash in slowly dripping tap water	3–5 min
(6)	Dehydrate or run slides up in the usual manner with 70% and 95% alcohol.	
(7)	Counterstain with a 95% alcoholic solution of eosin	about 1 min
(8)	Dehydrate with absolute alcohol in the usual manner.	
(9)	Clear and mount in the usual manner.	

d. Results.—

Nuclei—bright clear blue
Cytoplasmic and other tissue components—pink to red

2. Short Manual Procedure.—

A modification of the above procedure which saves time is as follows.

a. Procedure.—

(1)	Xylene, 2 changes Dip gently, not vigorously.	20 dips in each change
(2)	95% Alcohol, 2 changes	5 dips in each change
(3)	Removal of mercuric pigment if tissue is fixed in mercuric chloride-containing fixatives.	

Step	Procedure	Time
	(a) Lugol's solutions	30 sec
	(b) Distilled water	2 min
	(c) 5% Sodium thiosulfate	30 sec
(4)	If tissue was not fixed in a mercuric chloride-containing fixative, skip Steps (3) (a), (b), and (c), and wash in slowly dripping tap water	2 min
(5)	Hematoxylin stain	1 min
(6)	Wash in slowly dripping tap water	2 min
(7)	Blue the hematoxylin (except iron hematoxylin) in an alkaline solution, such as ammonia water.	
(8)	Wash in slowly dripping tap water	2 min
(9)	95% Alcohol	10 dips
(10)	Counterstain with a 95% alcoholic solution of eosin	1 min
(11)	Absolute alcohol, 3 changes	1 or 2 dips each
(12)	Xylene, 3 changes	1 or 2 dips each
(13)	Mount cover slip.	

b. **Results.**—

Nuclei—bright clear blue
Cytoplasmic and other tissue components—pink to red

3. **Automated Procedure.**—

Routine hematoxylin and eosin staining can be carried out automatically in an instrument such as the Mono-Autotechnicon. Other instruments which can perform automatic routine staining are the Ames' Histo-tek and Fisher's dual unit Tissuematon.

Operation of the Mono-Autotechnicon is basically similar to that of the Ultra-Autotechnicon, except that staining occurs without heat or vacuum. Specific directions are found in the manual of operations. Reagents used in the procedure could be those purchased from Technicon Corporation or those that are ordinarily used in manual procedures.

The steps and times used in this procedure have been modified from those suggested for staining in the Autotechnicon's manual, as it was found that the eosin ran out of the tissue during the dehydration steps for the times listed in the manual.

a. **Procedure.**—

		Time (min)
(1)	Deparaffinization, Paraway I (Technicon's Paraway, DP-9-50)	4
(2)	Hydration and removal of mercuric pigment (0.25% iodine in Technicon dehydrant, S-29)	2
(3)	Hydration in S-29	2
(4)	Distilled water	2
(5)	Nuclear staining (Ehrlich's hematoxylin or Technicon's hematoxylin solution)	5
(6)	Distilled water	4
(7)	Bluing of hematoxylin (0.01N Technicon's lithium carbonate)	2
(8)	Distilled water	2
(9)	Cytoplasmic counterstaining (95% alcoholic solution of eosin or Technicon's eosin solution)	5
(10)	Dehydration in S-29	1
(11)	Dehydration in S-29	1
(12)	Clearing, Paraway II	3
(13)	Clearing, Paraway III	3
(14)	Mount cover slip	
	Total Time:	36 min

b. **Results.**—

Nuclei—bright clear blue
Cytoplasmic and other tissue components—pink to red

c. **Notes.**—

(1) Beaker containing Paraway III is outside of the Autotechnicon and should be covered to avoid contamination and evaporation.

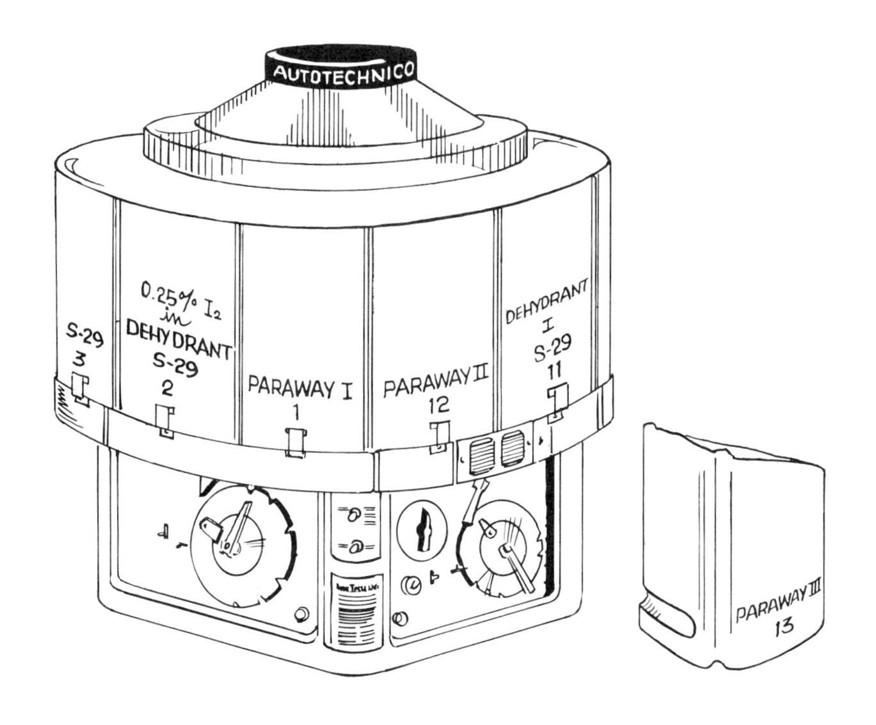

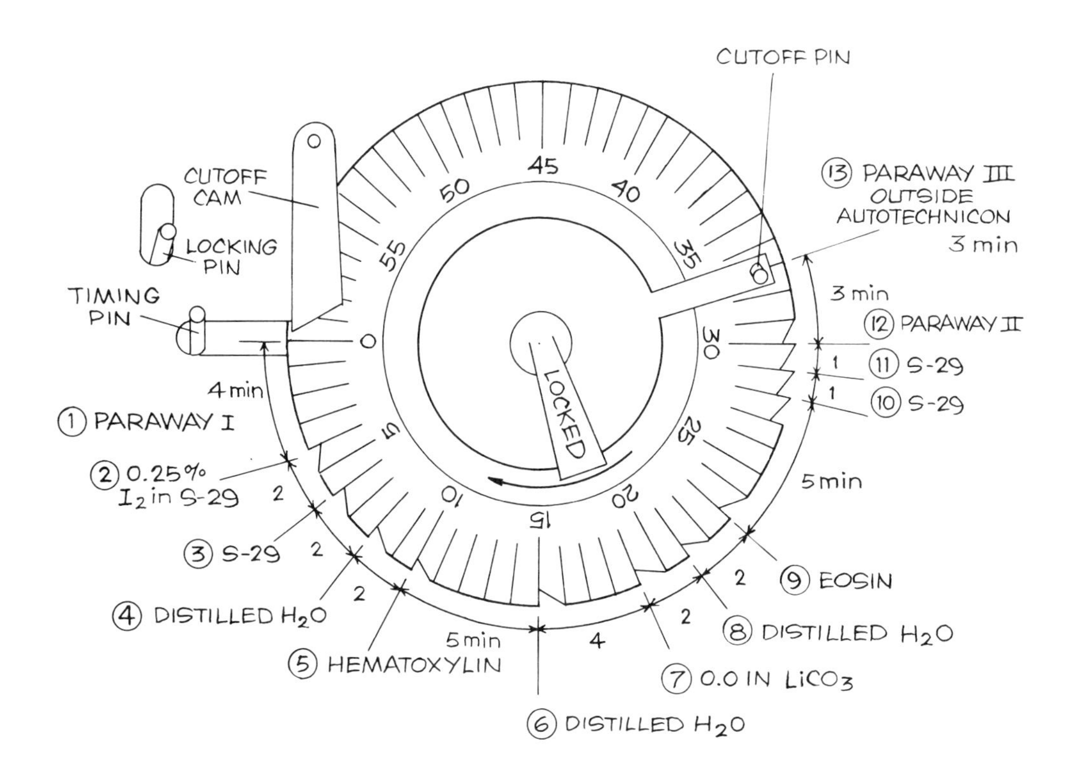

FIG. 7.3. MONO-AUTOTECHNICON FOR ROUTINE HEMATOXYLIN AND EOSIN STAINING

(2) Fill beakers to the lower staining mark (700 ml) on the beaker.

(3) Changing of solutions depends on volume of work. Technicon's suggested schedule is as follows:

Paraway—once a week
Technicon dehydrant—every 10–20 days
Distilled water—for every rack of slides
Technicon lithium carbonate—every week
The 0.01N lithium carbonate is prepared by dissolving one premeasured packet of this reagent in 1 liter of distilled water
Staining solutions—every 4 months with constant use; however, the authors suggest that these solutions be changed more frequently, especially the eosin which, if it loses its acidic nature, will not counterstain the tissue components for reasons mentioned before.

B. Regressive Method

In this method the tissue sections are first overstained and then destained until the correct intensity of stain is obtained. The destaining step is also called differentiation and can be accomplished by differentiating agents, such as acids, bases, oxidizing agents, alcohols and mordants.

a. Procedure.—

(1)	Deparaffinize and hydrate slide to water. Remove mercuric pigment if present.	
(2)	Overstain with a hematoxylin solution such as Delafield's or Harris' hematoxylin	15–20 min
(3)	Wash in slowly dripping tap water	3–5 min
(4)	Dehydrate in 70% alcohol	few dips
(5)	Differentiate in acid alcohol (0.5% HCl in 70% alcohol) for a few seconds. Blue the nuclei with alkaline (ammoniated) alcohol for a few seconds (70% alcohol with a drop of concentrated ammonia added). Examine under the microscope. If nuclei are still too dark repeat the differentiation.	
(6)	When correct intensity of the nuclei is obtained, thoroughly blue the nuclei in alkaline alcohol solution	3–5 min
(7)	Rinse in 70% alcohol	few dips

(8) Counterstain with a 95% alcoholic solution of eosin — about 1 min

(9) Dehydrate in absolute alcohol in the usual manner.

(10) Clear and mount.

b. Results.—

Nuclei—bright deep blue
Cytoplasmic and other tissue components—pink to red

QUESTIONS

Cover answers with a piece of paper. Answers appear at end of questions.

(1) The use of dyes or stains makes tissue components visible by
- (a) Staining only the nuclei
- (b) Staining only the cytoplasm
- (c) Bringing out differences in refractive indexes of tissue components
- (d) Ripening the cells

(2) One of the first and most widely used stains in histology is
- (a) Weigert's iron hematoxylin
- (b) Heidenhain's iron hematoxylin
- (c) Mallory's iron chloride hematoxylin
- (d) Hematoxylin

(3) The following stains: Delafield's hematoxylin, Ehrlich's hematoxylin, Harris' hematoxylin and Mayer's hematoxylin, have something in common besides hematoxylin, that is
- (a) They all contain ferric chloride
- (b) They all contain HCl
- (c) They all contain mercuric oxide
- (d) They are all "direct" hematoxylin stains

(4) "Mordant" hematoxylin stains are characterized by
- (a) The mordant precedes the hematoxylin
- (b) They all contain H_2SO_4
- (c) They are all variations of Mallory's iron chloride hematoxylin
- (d) They are all cytoplasmic stains

(5) The most widely used cytoplasmic stain for general purposes is
- (a) Acid fuchsin
- (b) Congo red
- (c) Biebrich scarlet
- (d) Eosin

(6) The best fixative to bring out the brilliance of the answer chosen in (5) preceding is
- (a) Bouin's
- (b) Formalin
- (c) Zenker's
- (d) Osmium tetroxide

(7) In the basic procedure for staining paraffin infiltrated tissues, a sequence is followed. In the following sequence (a)–(d), which is out of order?
(a) Deparaffinization
(b) Absolute alcohol
(c) Washing
(d) Hydration with decreasing concentrations of alcohols

(8) The process of "clearing" involves the use of
(a) Mordanting chemicals
(b) Xylene or toluene
(c) Increasing concentrations of alcohol
(d) (a) and (c) only

(9) In the following summary of the General Procedure for staining, one step is out of order. Which one?
(a) Deparaffinization
(b) Absolute alcohol
(c) Hydration
(d) Washing
(e) Staining and counterstaining
(f) Dehydration
(g) Clearing
(h) Mounting the cover slip
(i) Absolute alcohol

(10) Another term for deparaffinize is
(a) Hydrate
(b) Dehydrate
(c) Decerate
(d) Infiltrate

Answers

(1) c
(2) d
(3) d
(4) a
(5) d
(6) c
(7) c
(8) b
(9) i
(10) c

8

Mounting of the Cover Slip and Types of Mounting Media

The purpose of mounting a cover slip over the stained tissue sections is to make permanent preparations. The usual mounting medium or mountant for routine histotechnique is a resinous material soluble in highly volatile hydrocarbons such as xylene or toluene. In order to bring the tissue sections into a solvent similar to the mounting medium, the slides are run up a series of increasing concentrations of alcohols to xylene. The mounting medium is applied and cover slip placed over the tissue sections (see Exercise 7). As the solvent of the mounting medium evaporates, the resin cements the cover slip to the slide.

I. METHODS OF COVERSLIPPING

There are several methods of coverslipping. Only two will be described.

A. Forceps Method

1. Make sure mounting medium is of correct consistency. It should be liquid and should flow easily. It should not be viscous; if it is viscous add some solvent and stir.
2. Work on a clean flat surface.
3. Do not allow sections to dry out before cover slip is applied. Remove slides from the xylene one at a time.
4. Wipe excess stain and other debris off the slide with a gauze pad moistened with xylene.
5. Use a clean, correct size cover slip. The size of the cover slip used depends on the size and number of tissue sections to be covered. The cover slip should extend several millimeters beyond the margin of the tissue sections.
6. Apply a thin streak of the mounting medium to the cover slip. The quantity of mounting medium, which is determined by experience, must be sufficient to cover the space between cover slip and slide and fill the tissue and tissue spaces.

   The mounting medium should spread slightly beyond the edge of the cover slip, as shrinkage takes place upon drying. If amount of mounting medium is not sufficient, the shrinkage leaves the cover slip no support and air bubbles will also form, which can cause the stain to fade. If excess amount of mounting medium is used, sticky, unsightly slides result, which are hard to clean (see Section II following).

7. After streaking the cover slip, turn cover slip over and rest on edge on the slide close to the sections. Hold the other edge with a flat-jawed cover slip forceps.
8. Lower the cover slip into place slowly to allow air to escape from under the cover slip. After cover slip is over sections, press firmly but gently with back of forceps from center outward to distribute mounting medium evenly and to remove, if possible, any entrapped air bubbles.

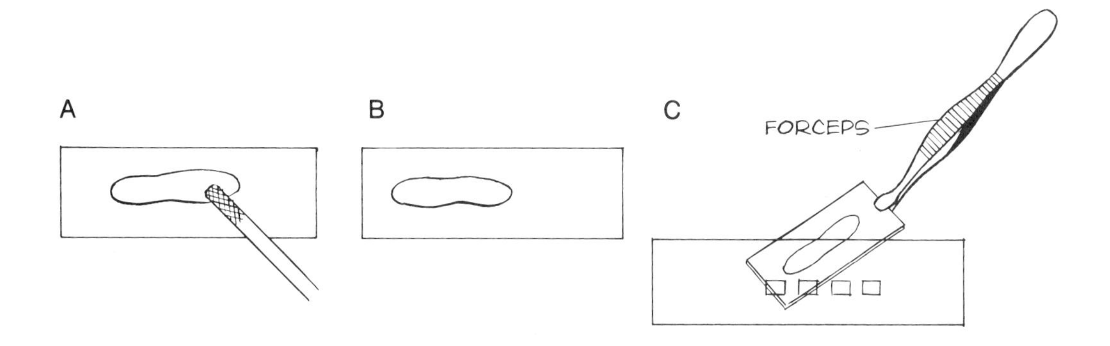

FIG. 8.1. FORCEPS METHOD OF MOUNTING A COVER SLIP.
A—Streaking cover slip.
B—Invert cover slip.
C—Lower cover slip with forceps over tissue sections.

B. Inverted Slide Method

Steps 1 to 5 as described in Section A apply to this method also.

6. Place a small round drop of mounting medium in the center lower edge of the cover slip.
7. Invert slide and bring slide up to edge of the cover slip so that the slide touches the mounting medium. Lower slide gradually over the cover slip to ease out the air. Press gently in place.

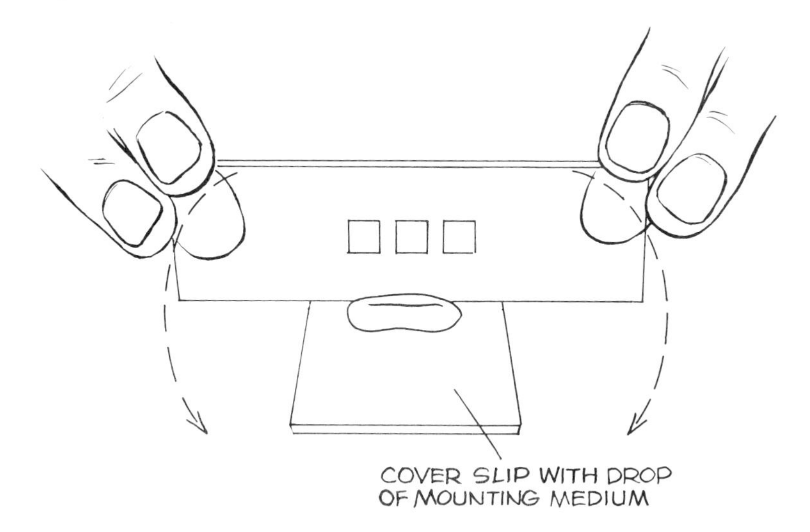

FIG. 8.2. INVERTED SLIDE METHOD

If air bubbles can not be easily removed, place slide into xylene until the cover slip drops off; rinse and remount the cover slip.

II. DRYING AND CLEANING

To prevent cover slip movement and damage to the tissue sections, slides must be kept horizontal until the mounting medium has hardened and cover slip is attached to the slide.

After cover slip is firmly attached, excess mounting medium that has oozed from under the cover slip can be scraped off with a razor blade and the remainder cautiously removed with a gauze pad or cotton-tipped wooden applicator moistened in xylene.

III. TYPES OF MOUNTING MEDIA

There are two types of mounting media, water-insoluble or resinous, and water-soluble or aqueous mounting media.

A. Water-insoluble or Resinous Mounting Media

The resinous mounting media may be either natural or synthetic and are soluble in highly volatile organic reagents, such as xylene or toluene.

1. **Natural Resinous Mounting Media.**—The natural resins are sticky materials expressed from evergreen trees, such as Canada balsam or gum damar. The natural resins have a series of disadvantages that render them less useful than the synthetic resins. The disadvantages are:

   a. Chemical composition is variable; it is a mixture of several organic chemicals
   b. Dry slowly—takes months for the cover slip to permanently attach to the slide
   c. Turn yellow with age
   d. Crack with age as the solvent evaporates
   e. Acidic in pH—cause fading of basic stains.

2. **Synthetic Resinous Mounting Media.**—These are made in the laboratory, they are β-pinene polymers. When purchased, the mounting media contain 60% resin dissolved in the solvent. The synthetic mounting media are superior to natural ones in several respects. The advantages are:

   a. Composition is known, stable and inert
   b. Media are readily soluble in volatile hydrocarbons
   c. Pale in color
   d. Do not yellow with age
   e. Dry within a reasonably short period of time
   f. Adhere tightly to the glass of the slide and the cover slip

g. Neutral in pH, so do not fade the stains
h. Have the same refractive index (R_f) as tissue and glass ($R_f = 1.53–1.54$)

This means that light passing through the glass, mounting medium and tissue section readily makes the tissue section transparent, but not to the extent that important tissue components become difficult to see.

The synthetic mounting media are sold under various commercial trade names, such as:

Permount
Histoclad
Bioloid
Technicon Mounting Medium
Kleermount and
Harleco

All are equally good.

A suggestion has been made that to prevent air bubbles under the cover slip one should have the stained tissue sections in the same solvent as that in which the resin is dissolved.

B. Water-soluble or Aqueous Mounting Media

Aqueous mounting media are used when tissue components or stains are soluble in alcohol or the hydrocarbons. Tissue sections are mounted directly from water. The components of aqueous mounting media are:

1. Gelatin or gum arabic (acacia)—acts as a solidifying agent
2. Sugar and salt to increase refractive index
3. Glycerol (glycerin)—keeps tissue soft and prevents cracking and drying of the tissue
4. A mold growth inhibitor such as phenol or thymol

A few examples of aqueous mounting media are Kaiser's glycerol jelly, Von Apathy's gum syrup and Farrant's Medium.

IV. LABELING THE SLIDES

After the slides have been cleaned and the cover slip firmly attached, the slides are ready to be labeled.

A. Place the slide so the specimen is properly oriented for later study.
B. Place a square label on the left side of the slide, on the same side as the cover slip.
C. Use waterproof ink and record the following information:

1. Institution in which the slide was prepared
2. Code number

3. Name of tissue or organ and type of section
4. Name of organism and sex
5. Type of fixative
6. Type of stain
7. Section thickness
8. Date on which the slide was completed

Abbreviate where possible. The information on the slide label should correspond exactly to the information recorded in the laboratory notebook for the processing of each particular tissue.

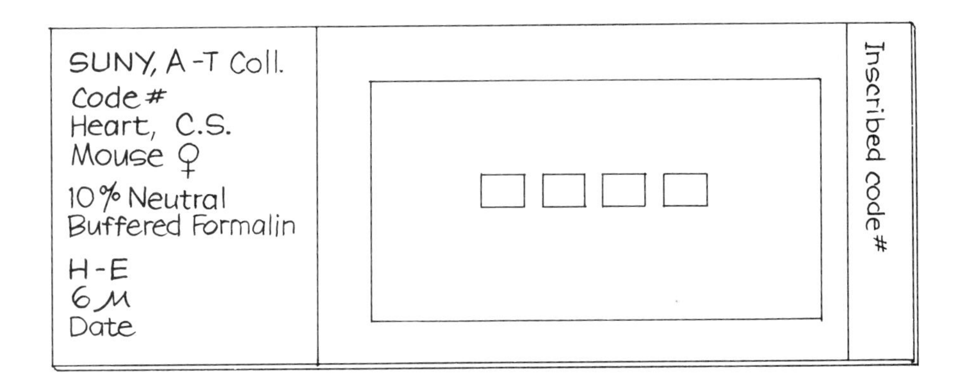

FIG. 8.3. FINISHED SLIDE

V. STORING AND CATALOGUING THE SLIDE IN THE SLIDE BOX REGISTER

The slide must be protected from breakage, light, moisture and dust. Histology slides are best stored in a slide box which has numbered slots. The boxes should be stored on end so the slides are horizontal.

In the slide register corresponding to the numbered slot, list the type of tissue or organ and type of section on the slide.

QUESTIONS

Cover answers with a piece of paper. Answers appear at end of questions.

(1) The most commonly used mountant is
- (a) Gelatin
- (b) Cedarwood oil
- (c) Glycerin
- (d) A resinous material soluble in volatile hydrocarbons

(2) If a mounting medium becomes too viscous, one should add
- (a) Solvent for the medium
- (b) 1% Acid-alcohol
- (c) H_2O plus Mayer's adhesive
- (d) Nothing

(3) Is the following statement true or false? "The mounting medium should spread slightly beyond the edge of the cover slip, as shrinkage takes place upon drying."
- (a) True
- (b) False
- (c) Partly true; partly false

(4) After cover slip is over sections, one should press firmly but gently from center outward to distribute mounting medium evenly and to remove entrapped air bubbles with
- (a) Fingernail
- (b) Jaws of cover slip forceps
- (c) Fingertip
- (d) Back of forceps

(5) If air bubbles can not be easily removed, one should
- (a) Make a new slide
- (b) Leave them in place
- (c) Heat to 80°C
- (d) Soak slide in xylene until cover slip drops off and remount the cover slip

(6) In cleaning excess mounting medium from a slide, it is best to use
- (a) A sharp scalpel and water
- (b) A razor blade and xylene
- (c) A wooden tip applicator with acid-alcohol
- (d) HCl—50%

(7) Which of the following is <u>not</u> an advantage of synthetic resinous mounting medium?
- (a) Composition is known, stable and inert
- (b) Pale in color
- (c) Dries in a reasonably short time
- (d) Acidic in pH

(8) Canada balsam and gum damar are
- (a) Excellent clearants
- (b) Great for "oil immersion"
- (c) Natural resinous mounting media
- (d) Expensive but good mordants

(9) Gelatin or acacia, sugar and salt, phenol or thymol, and glycerol are components of
(a) Natural resinous mounting media
(b) Synthetic resinous mounting media
(c) Aqueous mounting media
(d) Gum arabic

(10) Which of the following are not necessary on the slide label? If all are necessary, answer (i).
(a) Name of institution
(b) Code number
(c) Name of tissue or organ
(d) Type of section
(e) Name of organism and sex
(f) Type of fixative and stain
(g) Section thickness
(h) Date
(i) All the above are necessary

Answers

(1) d
(2) a
(3) a
(4) d
(5) d
(6) b
(7) d
(8) c
(9) c
(10) i

9

Theoretical Aspects of Staining

Tissue components do not have enough color and are usually transparent. The addition of chemical reagents known as dyes or stains imparts a characteristic color to each tissue component which increases its visibility.

This exercise is concerned with the chemical structure and action of the stains.

I. TYPES OF STAINS

Stains can be classified as natural or synthetic.

A. Natural Stains

The natural stains have been discussed in Exercise 7.

B. Synthetic Stains (Coal Tar or Aniline Dyes)

Synthetic stains are of all colors and are brilliant in appearance. They are synthetically produced from hydrocarbons by the distillation of coal tar. These stains are also known as aniline dyes, as aniline is an intermediate product in the preparation of a large number of them.

1. **Benzene Ring.**—The benzene ring

```
(       H       )
(       |       )
(      /C\      )
(  H-C     C-H  )
(    |     |    )
(  H-C     C-H  )
(      \C/      )
(       |       )
(       H       )
```

is the chemical basis upon

which the synthetic stain molecule is built. Benzene is not colored, but by replacing some of the hydrogens with specific chemical groupings (radicals) or by changing the molecular configuration of the ring, due to resonance, color is imparted to the benzene.

2. **Chromophore.**—Those radicals which give color are known as chromophores, such as C=C, C=O, C=S, C=N, N=N, N=O, $\overset{O}{\overset{|}{N}}$-O. The resonant form that imparts color to the benzene ring is the quinoid structure

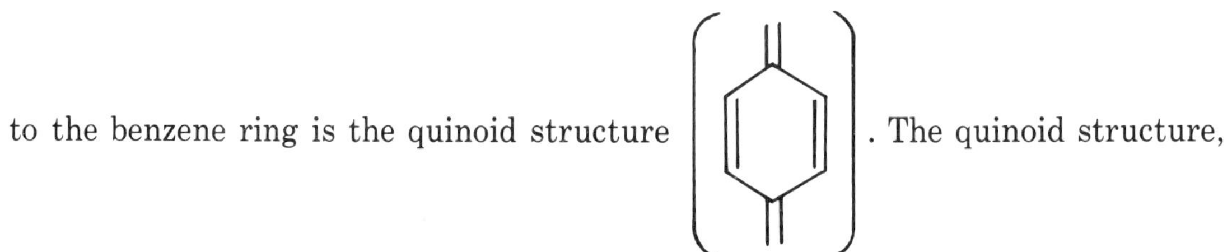

. The quinoid structure, the azo group (N=N), and the nitro group (NO_2) are the most common types of chromophores found in histological stains. When a chromophore is associated with the benzene ring, the compound acquires color and is called a chromogen, i.e.,

Benzene + Chromophore = Chromogen

3. **Auxochrome.**—The chromogen is not a tissue stain, as it has little attraction for the tissue components and can be easily removed. The chromogen must contain a group or groups that will allow it to become a salt and on solution electrolytically dissociate to become a charged substance, either an anion (negatively charged) or a cation (positively charged). The charge will allow the chromogen to be attracted to an oppositely charged group of the tissue components. The auxiliary group which is responsible for affinity of the chromogen for the chemical reagents of the tissue is known as an auxochrome. The replacement of one or more hydrogens of the benzene ring by an auxochrome makes the molecule an effective stain.

   Examples of auxochromes are amino ($-NH_2$), carboxyl (-COOH), hydroxyl (-OH) and sulfonic ($-SO_3H$) groups. In solution the amino groups are cationic or positively charged and have basic properties, while the latter three are anionic or negatively charged and have acidic properties. Whether a stain in solution has acidic (anionic) or basic (cationic) properties depends chiefly on its auxochrome and to a lesser extent on its ionization of the chromophores.

4. **Stain.**—A stain therefore is a benzene ring to which chromophore and auxochrome groups are attached. The color is given by the chromophore; the depth of color is reflected in the number of hydrogen atoms substituted on the benzene ring. If hydrogens are present, the stain tends to be yellow; as methyl or ethyl groups or larger complexes substitute for the hydrogens, the colors of the stains become red, violet, blue and green. The staining properties, i.e., the action of the stain with the tissue components, is due to the salt forming nature of the auxochrome, i.e.,

Benzene + Chromophore + Auxochrome = Stain

C. Acidic and Basic Stains

Stains as dry powders are stable salts.

An acidic or anionic stain is a salt of a colored acid, usually a salt of sodium.

$$Na^{+}\text{—Auxochrome}^{-}\text{—Chromogen}$$

A basic or cationic stain is a salt of a colored base, usually a salt of chloride.

$$\text{Chromogen—Auxochrome}^{+}\text{—}Cl^{-}$$

When these molecules are placed into solution they electrolytically dissociate to become acidic and basic stains.

The terms anionic and cationic are more appropriate than acidic and basic, since the stains themselves are not acids or bases but they possess the potential acidic or basic groups which give rise to these names. Thus, an acidic stain is one that has the ability to give up H^{+} or other positively charged groups, such as Na^{+}, to become negative or anionically charged. A basic stain is one that has the ability to give up OH^{-} or other negatively charged groups, such as Cl^{-}, to become positive or cationically charged.

D. Basophilic, Acidophilic and Neutrophilic Tissue Components

1. **Basophilic Tissue Components.**—Basic or positively charged stains have an attraction for acidic or negatively charged substances of the tissue, such as chromatin of the nuclei. These tissue components are considered the basophilic components, since they have a "love" for the basic stain.

2. **Acidophilic Tissue Components.**—The acidic or negatively charged stains have an attraction for basic or positively charged substances of the tissue, such as cytoplasm and connective tissue. These tissue components are considered the acidophilic components, since they have a "love" for the acidic stain.

3. **Neutrophilic Tissue Components.**—When basic and acidic stains are mixed and interchange ions, they form a new chemical compound or neutral stain. The neutral stain has attraction for certain tissue components which are considered neutrophilic components, such as the granules of the polymorphonuclear leukocytes.

II. NATURE OF STAINING ACTION

Staining reactions involve physical factors, chemical factors or a combination of both.

A. Physical Staining Action

Physical staining action is best explained by adsorption or condensation of the stain on the surface. Adsorption is a property of solid substances, such as the tissue components, to attract and to hold stains on their free surfaces. The adsorption of a stain concerns both physical and chemical factors. The physical factors are surface area, density and permeability of the adsorbing substances (tissue components) and size of the adsorbed particles (stain). In terms of size, the stain can have either molecular or colloidal dimensions. The chemical factor of adsorption is related to the electrostatic attraction of positive and negative groups on the surface of the tissue component between the ionized stain and the oppositely charged reactive groups. Thus, physical staining action is not only due to pure adsorption of the stain by physical factors to the surface, but also may involve a chemical reaction. One may then explain selective action of stains on a tissue by the fact that one stain acts on one part of the tissue by physical factors and another stain may act on another part of the tissue by chemical factors.

B. Chemical Staining Action

Chemical staining action assumes that the various chemical components of the tissue are amphoteric substances. An amphoteric substance behaves either cationically or anionically depending on the pH of the environment and will attract oppositely charged stain molecules (ions).

In chemical staining action there is first diffusion or penetration of the ionized stain into the chemical component of the tissue and then absorption of the stain. During absorption, the stain passes from the staining solution into the chemicals of the tissue component being stained, mainly proteins and nucleic acids. The reaction that occurs is in accord with the chemical law of ion exchange to form an ionic or electrostatic chemical bond. The actual chemical staining mechanism will be amplified shortly (Section III B).

C. Combination of Physical and Chemical Staining Action

Some staining action is a combination of physical and chemical staining. In combination staining action the stain is selectively adsorbed (physical staining action) onto the surface of components of tissue. Then, later, the stain penetrates into the component and becomes absorbed, i.e., chemically reacts with the chemicals of the tissue. Since chemical staining action is much more prevalent than physical staining action in routine staining procedures, the following discussion describes the mechanism of chemical staining.

III. CHEMICAL BASIS OF HISTOLOGICAL STAINING

A. Chemical Staining Properties

Chemical staining is influenced by several factors.

1. Concentration of stain, amount of ionization of the stain and amount of dissolved salt in the stain solution.—

With increasing concentration and ionization of the stain, the greater amount of stain becomes bound to those tissue components that accept the stain. If the amount of dissolved salt is increased in the stain, this will decrease staining action of a stain, as the salt ions compete with the ionized stain for reactive sites on the tissue chemical components.

2. Amount of ionization of the tissue chemical components.

3. pH of the environment in which the tissue chemical components and stain are found.—

   As will be amplified shortly, the pH of the environment influences the degree of ionization of tissue chemical components and stain.

4. Fixation treatment of the tissues.—

   Fixation may cause reorganization of the tissue chemical components to render certain chemicals more accessible to the stain molecules, thus increasing staining affinity. Or, fixation might change the permeability of the tissue components, allowing certain stains to have increased or decreased rates of penetration. Furthermore, the fixative might bind to certain chemical components of the tissue and leave other components free to accept the stain. For example, formalin fixation increases affinity of the chemical components of the tissue for the basic stain, as formalin combines to the amino groups of the tissue, leaving carboxyl groups free. Heavy metal fixatives, such as mercuric chloride fixatives, increase the affinity of the tissue for acidic stains, as these fixatives combine with carboxyl groups, leaving the amino groups free.

5. Temperature.—

   Increasing temperature increases the rate of penetration of the stain and staining action.

B. Theory of Chemical Staining Action

The theory of chemical staining assumes that the chemical components of the tissue are amphoteric in character. Depending on the pH of the environment, they behave as either acidic or basic substances and attract to themselves oppositely charged stain ions according to the chemical law of ion exchange to form ionic or electrostatic chemical bonds. To review, if the environment is acidic, the tissue chemical component acts as a basic or cationic substance; if the environment is basic, the tissue chemical component acts as an acidic or anionic substance (see Section II B).

For each chemical component of the tissue, especially the proteins and nucleic acids, there is a pH of the environment at which each molecule has an equal number of positive and negative charges. This pH is termed the isoelectric point (IEP). At the IEP there is a net algebraic charge of zero on the molecule. The molecule itself is known as the zwitterion. Since the zwitterion has zero or neutral charge on itself, it has little affinity or attraction for a charged stain molecule.

By adjusting the pH of the environment to the acidic side or below the IEP of the tissue chemical components, the free negatively charged carboxyl groups (COO^-) accept hydrogen ions (H^+) of the environment. This confers upon the molecule a net positive

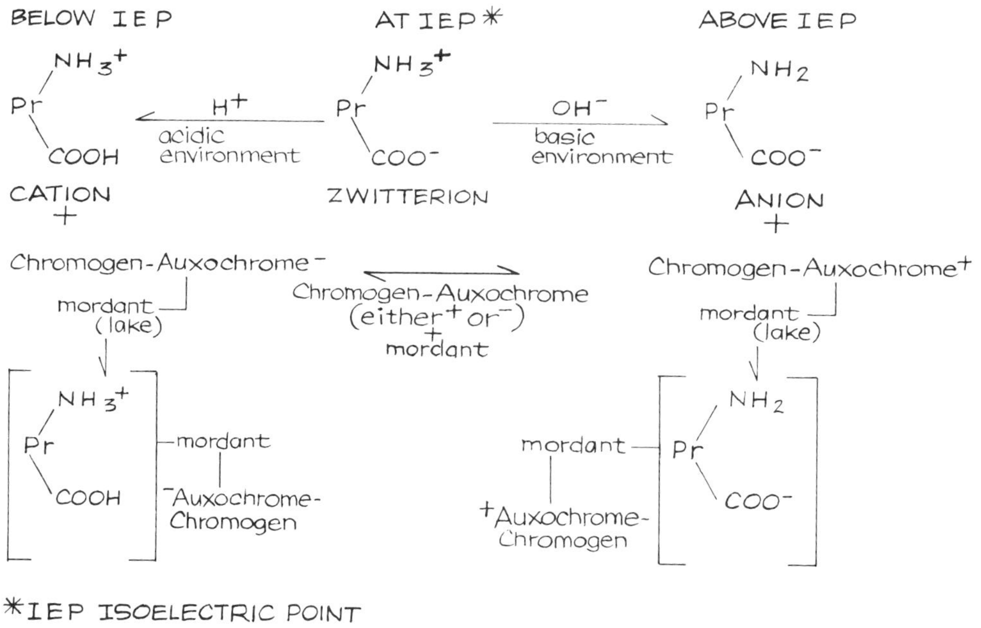

BELOW IEP — AT IEP* — ABOVE IEP

$Pr(NH_3^+)(COOH)$ ← H^+ acidic environment — $Pr(NH_3^+)(COO^-)$ — OH^- basic environment → $Pr(NH_2)(COO^-)$

CATION + — ZWITTERION — ANION +

Chromogen-Auxochrome⁻ ⇄ Chromogen-Auxochrome (either⁺ or⁻) + mordant ⇄ Chromogen-Auxochrome⁺

mordant (lake) — mordant (lake)

[$Pr(NH_3^+)(COOH)$]—mordant—⁻Auxochrome-Chromogen

mordant—[$Pr(NH_2)(COO^-)$]; mordant—⁺Auxochrome-Chromogen

*IEP ISOELECTRIC POINT

FIG. 9.1. EFFECT OF pH ON THE IONIZATION OF PROTEIN COMPONENTS OF THE TISSUE AND ACTION OF THE STAIN

charge, as there is a relative increase in number of free positively charged amino groups (NH_3^+), which can attract acidic or anionically (−) charged stain molecules to form an ionic bond.

If the pH of the environment is adjusted above the IEP of the tissue chemical components, i.e., on the basic side, the H^+ ions of the NH_3^+ group are lost to the environment to neutralize the basic conditions. This leaves the molecules with net negative charges, as it increases the relative number of the free COO^- groups, which can attract basic or cationically (+) charged stain molecules to form an ionic bond.

During routine hematoxylin and eosin staining, the pH of the hematoxylin stain is approximately 7, which allows the nuclear chemical components to ionize, giving up H^+. These components are left with a net negative charge, which will attract basic or cationically (+) charged stains. The pH of the eosin stain is slightly acidic, which allows the remaining chemical components of the tissue to accept H^+ from the environment. These components are left with a net positive charge, which will attract acidic or

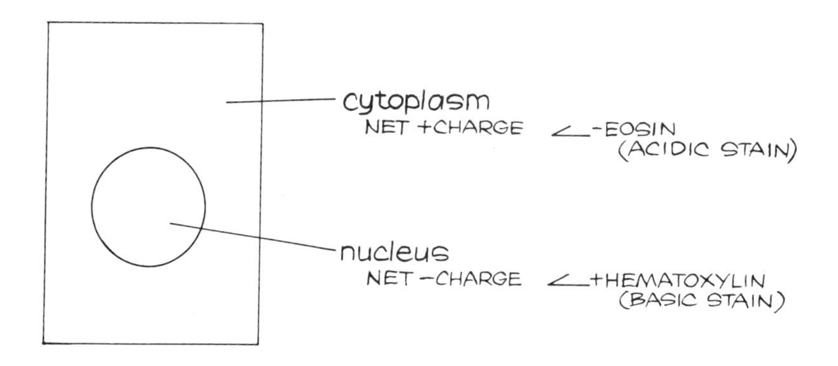

FIG. 9.2. AFFINITY OF HEMATOXYLIN FOR NUCLEUS AND OF EOSIN FOR THE CYTOPLASM

anionically (−) charged stains. This suggests that the IEP of the chemical components of the nucleus is on the acidic side of neutrality, while the IEP of the remaining chemical components of a tissue lies on the basic side of neutrality.

In addition to ionic bonding, hydrogen bonding and van der Waal's forces might also be involved in the chemical basis of staining.

C. Mordants

Up to now we have discussed the direct staining of the tissue components with the stain. Under this condition, the stain can be readily removed from the tissue component. To increase permanence of the stain for the tissue component an intermediary molecule can be interposed between the stain and this tissue component. This molecule is called a mordant and the staining is called indirect staining.

Mordants chemically are salts of divalent or trivalent metals, such as aluminum, iron or chromium. An important class of mordants is alums, which are salts of aluminum, such as ammonium alum ($NH_4Al(SO_4)_2 \cdot 12H_2O$). The stain-mordant combination forms a complex known as a lake.

The mordant acts by chemically bonding the stain to the tissue. It has been suggested that although electrostatic attraction draws the lake toward the charged chemical component of the tissue, it is probable that covalent chemical bonds are established between the mordant and the tissue chemical component rather than ionic bonds, otherwise the lake would not be held any tighter than the stain itself.

The advantage of mordanting is to make the stain relatively permanent, as the stain-mordant-tissue complex is insoluble in neutral aqueous and alcoholic solutions. This permits subsequent staining activity without removal of the first stain and allows for dehydration without decolorization of the stains.

Mordanting procedure can precede the stain, used either with the fixative or just prior to staining. Or the mordant can be used with the stain or subsequent to staining (see Section III D 2 on Regressive Staining).

D. Progressive and Regressive Staining

1. **Progressive Staining.**—In progressive staining the intensity of color of the tissue components becomes greater with time, until the desired color is reached. During progressive staining the mordant is used preceding the stain or with the stain.

2. **Regressive Staining.**—A regressive staining procedure allows the tissues to be first overstained and then partly destained or decolorized in order to differentiate the various tissue components from each other. The amount of differentiation is controlled under the microscope. Differentiating agents are of various types such as water, alcohol, dilute acids and bases, oxidizing agents, mordanting solutions, or the stains themselves. The various differentiating agents act in different ways in order to destain the tissue.

E. Accentuators and Accelerators

1. **Accentuators.**—Accentuators are not mordants as they do not form lakes with the stain, but when employed with a stain they increase the intensity and selectivity of the stain.

2. **Accelerators.**—Accelerators are not mordants as they too do not form lakes with the stain, but when used with a stain they increase the rate of staining action.

F. Types of Chromasia

1. **Orthochromasia.**—Orthochromatic stains color the tissue components with the true color of the stain. The action of the stain is predictable with regard to color. If the stain is blue, it stains the tissue components blue and so forth. The majority of stains is orthochromatic. The process is called orthochromasia.

2. **Metachromasia.**—Metachromatic stains have the property of staining tissue components in colors different from that of the stains themselves. The process is called metachromasia. The tissue components that cause change in color of the stain are called the chromotropes.

   Metachromatic stains do not chemically form a specific group, but are distributed among the different chemical groups. The color change of the metachromatic stains is from blue through purple to red. The metachromasia is thought to be caused by the chromotrope in the tissue promoting polymerization of the stain molecules. The chromotrope, itself, must be a macromolecule with numerous reactive sites on its surface so as to bind various amounts of stain in such a way as to cause interaction among the stain molecules thus causing the color shift.

   Tissue components which are metachromatic are cartilage, mucus, and mast cell granules. These components contain negatively charged sulfate acidic polysaccharides, termed mucopolysaccharides.

   The metachromatic stains, for example, thionin or toluidine blue, are basic stains, normally blue in color, and change colors to purple through red depending upon the degree of polymerization that takes place. Metachromatic staining techniques are significant, as they may give an indication of the chemical composition of the tissue component accepting the stain.

3. **Polychromasia.**—Polychromatic stain is a mixture of several stains; each stain retains its own color. The original stain is a single stain which spontaneously forms into other stains. The process is called polychromasia. An example of a polychromatic stain is methylene blue. Some of the methylene blue on standing spontaneously oxidizes to azure A and B and methylene violet. The polychromatic stain is a mixture of these stains. Blood cell stains are polychromatic stains.

4. **Allochromasia.**—Allochromatic stains are not stable in solution and give rise to other color molecules which are impurities. The process is called allochromasia. Allochromasia is observed when an orthochromatic stain breaks down to other stains, which then stain the tissue a different color from the original orthochromatic stain. This is also considered to be false metachromasia due to the stain impurity.

QUESTIONS

Cover answers with a piece of paper. Answers appear at end of questions.

(1) Radicals such as C=C, C=O, C=S, N=N, etc., which provide color are known as
(a) Benzene ring
(b) Chromogens
(c) Chromophores
(d) Chrominoids

(2) An auxochrome is
(a) The same as a quinoid
(b) The auxiliary group(s) which is responsible for affinity of the chromogen for chemical reagents of the tissue
(c) A replacement for a hydrogen of the benzene ring
(d) A replacement for a chromophore

(3) Is the following statement true or false? "A stain is a benzene ring to which chromophore and auxochrome groups are attached."
(a) True
(b) False

(4) Chromatin of nuclei is
(a) Negatively charged and attracts basic stains
(b) Positively charged and attracts acid stains
(c) Neutral and accepts basic and acidic stains
(d) Neutral and accepts neither basic nor acidic stains.

(5) At the IEP there is a net algebraic charge of zero on the molecule, thus
(a) It attracts acidic stains
(b) Accepts basic stains
(c) Accepts acidic and basic stains
(d) Has little attraction for charged stain molecules

(6) The adsorption of a stain on the surface of a tissue is due to
(a) Chemical factors
(b) Physical factors
(c) Chemical and physical factors
(d) Neither chemical nor physical factors

(7) Which of the following does not influence chemical staining?
If all factors affect staining the answer is (e).
(a) Concentration of stain
(b) Amount of ionization of tissue chemical components
(c) Fixation treatment of the tissue
(d) Temperature
(e) All of the above affect staining

(8) Formalin fixation increases affinity of the chemical components of the tissue for the basic stain since
(a) It contains heavy metals
(b) It combines with amino groups of the tissue
(c) It combines with carboxyl groups
(d) It causes tissue to be amphoteric

(9) To increase permanence of a stain for the tissue component, an intermediary molecule can be interposed between the stain and the tissue component. This molecule is called
(a) An ionizer
(b) Van der Waal force
(c) An IEP
(d) A mordant

(10) A "lake" is
(a) A stain-mordant combination
(b) An electrostatic force
(c) An alum compound
(d) Temporary stain for frozen sections

Answers

(1) c
(2) b
(3) a
(4) a
(5) d
(6) c
(7) e
(8) b
(9) d
(10) a

10

Specific Staining Procedures

In this exercise several examples of specific stains are presented so as to demonstrate to the student procedures involved in differentiating and visualizing specific tissue components. Extensive routine and specific staining methods can be obtained from other histotechnique manuals, one such example from a list of many is *Manual of Histologic Staining Method of the Armed Forces Institute of Pathology, 3rd Edition.* 1968. L. Luna (Editor). McGraw-Hill Book Co., New York.

I. CONNECTIVE TISSUE FIBERS

By the application of specific stains, collagenous and elastic fibers can be demonstrated.

A. Collagenous Fibers

There are two standard methods for the demonstration of collagenous fibers; these are Mallory and Masson's staining procedures. Although these staining procedures are considered multiple stains, each stain reacts orthochromatically.

The Mallory triple stain consists of acid fuchsin and a mixture of aniline blue and orange G. The acid fuchsin, which is a cytoplasmic stain, is used first, followed by collagenous fiber staining employing phosphomolybdic acid and/or phosphotungstic acid or oxalic acid as a mordant for aniline blue.

The exact mechanism of how the acids act as mordants is not known, but it is suggested that since these acids have relatively more acidic groups than the collagenous fibers have basic groups, these acids, after being bound to the basic groups of the collagenous fibers, are left with free acidic groups that can bind to other basic groups. Aniline blue is an amphoteric stain, and ordinarily behaves as an acidic acid, but after treatment of the tissues with phosphomolybdic, phosphotungstic or oxalic acid, the stain behaves under these pH conditions as a basic stain, being attracted to the acidic groups on these acids. In this way aniline blue selectively stains the collagenous fibers. Without these acids attaching to the collagenous fibers, aniline blue would have very little selectivity.

The acid fuchsin, even though an acidic stain, does not remain in the collagenous fibers to stain them. One reason that has been suggested to explain this is the pH of acid fuchsin may not be acidic enough to be attracted sufficiently to the acidophilic collagenous fibers. Since no chemical reaction takes place, the stain will diffuse out on placing the tissue into the phosphomolybdic, phosphotungstic or oxalic acids. A purely physical reason is based on the looser texture of collagenous fibers compared to that of cytoplasm; this results in different permeabilities of the tissue components. The collagenous fibers having a greater permeability would promote stain diffusion unless a mordant were initially present to bind the stain. Thus, as the phosphomolybdic, phosphotungstic or oxalic acids diffuse into the collagenous fibers, they "push" out the acid fuchsin and they, being sufficiently acidic, in turn become bound to the collagenous fibers enabling these fibers to be stained by the aniline blue for the aforementioned reason.

The Mallory method is good not only for demonstrating collagenous fibers but also for the general topography of the tissues, as the third stain, the acidic orange G stains the same acidophilic tissue components orange to yellow, while the aniline blue, acting as an acidic stain, stains the cartilage matrix and mucus blue.

The method is not useful in differentiating cellular structures. Since nuclei stain poorly a red to pink with acid fuchsin, it is possible to precede this method by a basic staining procedure in order to stain the nuclei more intensely.

1. Triple Stain Procedures.—

a. Mallory's Stain—Pantin Method.—(Humason 1972)

(1) *Fixation.*—Any general or routine fixatives; Zenker's type recommended

(2) *Section.*—6–10 μ thick

(3) *Solutions.*—

1% Acid Fuchsin (Mallory I Solution)

Acid fuchsin	1.0 g
Distilled water	100.0 ml

The acid mordants can be either a 1% solution of phosphomolybdic, phosphotungstic or oxalic acid. Phosphomolybdic acid reacts more specifically to collagenous fibers, while the phosphotungstic acid intensifies plasma staining. Oxalic acid lowers the pH and makes the aniline blue stain more rapidly and intensely.

1% Phosphomolybdic Acid

Phosphomolybdic acid	1.0 g
Distilled water	100.0 ml

Aniline-orange G (Mallory II Solution)

Aniline blue, WS (methyl blue)	0.5 g
Orange G	2.0 g
Distilled water	100.0 ml

An alternate Mallory II Solution:

Solution A

2% Aniline blue, WS

Solution B

1% Orange G
Distilled water

Solution C

1% Phosphomolybdic acid

Keep these solutions separate until use, since mixture may decompose. Mix equal parts of each just before use.

Sirius supra-blue stain can be used in place of Mallory II, but do not mix this stain with phosphomolybdic acid.

Sirius supra-blue[1]	2.0 g
Distilled water	100.0 ml
Glacial acetic acid	2.0 ml

[See staining procedure for use of this stain (p. 102)]

(4) *Procedure.—*

(a)	Decerate and hydrate tissue sections to water; remove mercuric pigment if present (see p. 70). Rinse in distilled water. If mercuric chloride is absent from the fixative, mordant with saturated mercuric chloride (about 7%) dissolved in 5% acetic acid for 10 minutes. Wash and treat tissue sections to remove excess mercuric pigment as above.	
(b)	(Optional Step) Stain nuclei in hematoxylin as in a routine hematoxylin stain	
(c)	Stain in Mallory I solution	15 sec
(d)	Rinse in distilled water to wash out excessive acid fuchsin	about 10 sec
(e)	Mordant with an acid mordant such as 1% phosphomolybdic acid	1–5 min

[1]Roboz Surgical Instrument Co., Washington, D.C.

(f)	Rinse in distilled water rapidly.	
(g)	Stain in Mallory II	1–2 min or less
(h)	Rinse in distilled water briefly.	
(i)	(Optional Step) 1% Acetic acid rinse, which contributes to transparency of the sections	1–2 min
(j)	Transfer to 90% alcohol to differentiate the aniline blue left in the tissue. Check under the microscope to see that the aniline blue is only in those tissue components as listed in the results following [(5) Results]. The tissue sections in general should have a muddy purple appearance.	
(k)	Dehydrate in absolute alcohol, 2 changes After the tissue is in the absolute alcohol, the tissue should change from muddy purple to clear blue and red.	a few dips in each
(l)	Clear and mount with a cover slip.	

If Sirius supra-blue stain is used instead of Mallory II, after Step (d) continue as follows:

(e)	Mordant with 1% phosphomolybdic acid	5 min
(f)	Wash with distilled water, 3 changes	10 sec in each
(g)	Stain with Sirius supra-blue	5 min
(h)	Wash in distilled water, 3 changes	10 sec in each
(i)	Dehydrate, clear and mount a cover slip as in Steps (k) and (l) preceding.	

(5) *Results.*—Using Mallory I and II solutions

Collagenous fibers—blue
Bone and cartilage matrix and mucin—blue
Erythrocytes, myelin—orange to yellow
Muscle, cytoplasm, nuclei—red to pink
Elastic fibers—pink or yellow

Modification of the Mallory's triple stain and the Masson's trichrome procedure is development of the "one step" stains.

b. **Mallory-Heidenhain Rapid One Step Method.**—(Emmel and Cowdry 1964). This staining method combines the several staining and mordanting steps into one staining step.

(1) *Fixation.*—Any general fixative

(2) *Section.*—5–10 μ thick

(3) *Solution.*—

Mallory-Heidenhain Stain

Dissolve in the following order.

Phosphomolybdic acid	1.0 g
Orange G	2.0 g
Aniline blue, WS	1.0 g
Acid fuchsin	3.0 g
Distilled water	200.0 ml

Solution is stable for several months. The authors have substituted phosphotungstic acid for phosphomolybdic acid as Humason (1972, p. 174) states that phosphomolybdic acid enhances fiber staining, while phosphotungstic acid intensifies plasma staining.

(4) *Procedure.*—

(a) Deparaffinize and hydrate tissue sections to water, remove mercuric pigment, if necessary.

(b) Mallory-Heidenhain stain — 5 min, but 1 min may be sufficient

(c) Wash in slowly running tap water.

(d) Dehydrate rapidly in 95% and absolute alcohols.

(e) Clear and mount a cover slip.

(5) *Results.*—

Collagenous fibers—blue
Bone and cartilage matrix and mucin—blue
Erythrocytes, myelin—yellow
Cytoplasm and nuclei of cells—red
Elastic fibers—pink or yellow

c. **Gomori's Rapid One Step Trichrome Stain.**—(Humason 1972)

(1) *Fixation.*—Any general fixative. Formalin fixed tissue should be mordanted overnight with a saturated solution of mercuric chloride.

(2) *Section.*—5–7 μ thick

(3) *Solutions.*—

Modified Bouin's Solution

Picric acid, saturated aqueous solution	15.0 ml (1 part)
Formalin, concentrated	75.0 ml (5 parts)
Glacial acetic acid	15.0 ml (1 part)
	105.0 ml

Mix just before use.

Trichrome Solution (Gomori's Stain)

Chromotrope 2R (C.I. 16570)	0.6 g
Aniline blue, WS	0.6 g
(or light green, SF [yellowish] or fast green, FCF)[2]	
Phosphomolybdic acid	1.0 g
(or phosphotungstic acid)	0.8 g[3]
Distilled water	100.0 ml
Then add hydrochloric acid, concentrated[4]	1.0 ml

Age for 24 hr in refrigerator before use. Store in refrigerator and use cold. Do not filter. Solution is stable until red stain begins to fade.

0.2% Glacial Acetic Acid

Glacial acetic acid	0.2 ml
Distilled water	100.0 ml

(4) *Procedure.*—

(a) Decerate and hydrate tissue sections to water; remove mercuric pigment if present.

[2]Depending upon color desired for the collagenous fibers, aniline blue, light green or fast green can be used as the fiber stain.
[3]Phosphomolybdic acid favors green-blue coloration, while phosphotungstic acid favors red coloration.
[4]Hydrochloric acid is substituted for glacial acetic acid as the pH is lowered to the point best for staining the collagenous fibers.

(b)	Treat with modified Bouin's solution, 56°C After use, discard. This step increases the affinity of the fiber stain for the collagenous fibers and interferes with cytoplasmic staining.	1 hr
(c)	Wash in running water, or until yellow color disappears from the tissue sections.	5 min
(d)	Stain in the Gomori's solution. Short staining time produces more red; longer staining time makes tissue stain green and blue.	about 1 min
(e)	Rinse briefly in distilled water to remove excess red stain.	
(f)	Rinse in 0.2% glacial acetic acid. This step makes the colors transparent without a change in the color.	30 sec
(g)	Dehydrate in 95% alcohol, 2 changes Do not remove too much of the green.	few dips in each
(h)	Dehydrate in absolute alcohol, 2 or 3 changes	few dips in each
(i)	Clear and mount a cover slip.	

Hematoxylin is not used to stain the nuclei, as most of the hematoxylin is removed after staining with the acidic trichrome stain.

(5) *Results.*—

Collagenous fibers—blue or green
Muscle fibers—red, striations easily seen
Other tissue components—red

2. **Picro-ponceau with Hematoxylin.**—(Humason 1972). As discussed above, the affinity of collagenous fibers for acidic stains depends on strongly acidic solutions. Picric acid provides the acidic pH for the selective staining of the collagenous fibers. If the pH rises, staining of the collagenous fibers will not occur. Furthermore, picric acid must be a saturated solution, otherwise the collagenous fibers stain pale pink to orange rather than light red. Picric acid also serves as a counterstain, staining the cell's cytoplasm yellow. The picro-ponceau stain replaces the picro-acid fuchsin stain (Van Gieson's stain), as the latter stain rapidly fades, but the colors are identical.

(1) *Fixation.*—Any general fixatiave

(2) *Section.*—5–10 μ thick

(3) *Solutions.*—

Hematoxylin (see Exercise 7)

Picro-ponceau Stain

Ponceau S (C.I. 27195), 1% aqueous	1.0 ml
Picric acid, saturated aqueous	86.0 ml
Acetic acid, 1% aqueous	4.0 ml

(4) *Procedure.*—

(a)	Deparaffinize and hydrate tissue sections to water; remove mercuric pigment if necessary.	
(b)	Stain in hematoxylin Weigert's iron hematoxylin is the suggested nuclear stain as the iron mordant is already present in the staining solution. If another hematoxylin is used, mordant first with iron alum mordanting for 5–10 min, wash in running tap water, and then stain with hematoxylin. Prolonged hematoxylin staining and iron mordanting help to prevent picric acid from decolorizing the nuclear stain too much.[5]	10–15 min
(c)	Wash in running tap water until nuclei are deep blue to black depending on the hematoxylin used.	
(d)	Picro-ponceau counterstain Rinse in distilled water Check under the microscope for sharpness of the nuclei. Continue to stain in the counterstain and destain or differentiate in distilled water until nuclei are sharp and distinct. If too much hematoxylin was extracted, return to the nuclear stain.	3–5 min few seconds
(e)	Dehydrate in 70% alcohol	several dips

[5] Another method to prevent fading of hematoxylin during picro-ponceau staining is to heat tissue with a 1:1 solution of 2.5% aqueous phosphotungstic and phosphomolybdic acids for 0.5–1.0 min before staining in the counterstain. After this treatment, wash briefly to remove excess acids and proceed to picro-ponceau stain.

(f) Dehydrate in 95% alcohol, 2 changes — several dips in each

(g) Dehydrate in absolute alcohol.

(h) Clear and mount a cover slip.

(5) *Results.*—

Collagenous fibers—red
Nuclei—blue to black according to stain used
Other tissue components—yellow

B. Elastic Tissue

Elastic tissue can be demonstrated by the Verhoeff's stain and its modification, and by orcein. Elastic tissue occurs in the walls of blood vessels, such as larger arteries (aorta) and veins, in the form of sheets; and in the elastic cartilage, alveolar wall of lungs and in certain ligaments as fibers.

1. **Verhoeff's Elastic Tissue Stain.**—(Humason 1972). In this staining procedure the entire tissue section is overstained with Verhoeff's stain (hematoxylin-ferric chloride iodine lake). This complex diffuses into the elastic tissue and precipitates in it. The elastic tissue is differentiated from the other tissue components by removal of excess Verhoeff's stain with a ferric chloride mordant. Since the amount of mordant in the tissue is small compared to the quantity of mordant in the differentiating solution, the stain will be removed from those tissue components that have least affinity for it and form a soluble lake with the ferric chloride. The elastic tissue retains the stain the longest during differentiation because it has great affinity for the stain. The elastic tissue is stained black, and if differentiation is properly controlled, the other tissue components are red to orange in color.

(1) *Fixation.*—Any general fixative, but formalin or Zenker's type fixatives preferred.

(2) *Section.*—5–10 μ thick

(3) *Solutions.*—

Verhoeff's Working Stain Solution

Mix stock solutions.

A—Alcoholic hematoxylin	60.0 ml
B—10% Ferric chloride	24.0 ml
C—Verhoeff's iodine	24.0 ml

Stain is useful for 1–2 weeks. If elastic tissue stains poorly the staining solution has deteriorated.

Stock Solutions

A—Alcoholic Hematoxylin Solution

Hematoxylin	3.0 g
Absolute ethyl alcohol	60.0 ml

Dissolve hematoxylin into the alcohol by gentle heat of an electric hot plate, cool and filter.

B—Ferric Chloride Solution, 10%

Ferric chloride ($FeCl_3$)	10.0 g
Distilled water	100.0 ml

C—Verhoeff's Iodine Solution

Potassium iodide	4.0 g
Dissolve in distilled water	100.0 ml
Then add iodine	2.0 g

Differentiating Solution, Ferric Chloride, 2%

10% Ferric chloride	20.0 ml
Distilled water	80.0 ml

Picro-ponceau Stain (see p. 106)
or Van Gieson's Stain

Acid fuchsin, 1% aqueous	5.0 ml
Picric acid, saturated aqueous	100.0 ml

(4) *Procedure.—*

(a) Deparaffinize and hydrate tissue sections to water. Removal of mercuric pigment is not necessary for Zenker's fixed tissue as iodine in the Verhoeff's stain removes the mercury.

(b) Stain in Verhoeff's solution — 15 min–1 hr
Check under the microscope at 15 min intervals until tissue is black. Rinse excess stain with a few dips in distilled water. Hold slides in water while examining each slide under the microscope.

(c) When tissue is black, rinse excess stain in distilled water — few dips

(d)	Differentiate in 2% ferric chloride Rinse sections in tap water to stop the differentiation and check under the microscope for amount of differentiation. The differentiation should continue until the elastic tissue is sharp black and background is grey-black. Reverse the slides in the slide holder to ensure equal differentiation. If differentiation has proceeded too far, return tissue sections to Verhoeff's stain for another 5–10 min to restain, then differentiate again. This restaining and differentiation can continue as long as the tissue has not been treated with alcohol.	a few minutes
(e)	When differentiation is complete, rinse in distilled water	a few dips
(f)	Decolorize the iodine by placing tissue in 5% sodium thiosulfate	1 min
(g)	Wash in slowly running tap water	5–10 min
(h)	Counterstain in picro-ponceau or Van Gieson's stain Too long in the counterstain will destain the elastic tissue.	1 min
(i)	Differentiate and dehydrate in 95% alcohol, 2 changes	a few seconds in each
(j)	Dehydrate in absolute alcohol.	
(k)	Clear and mount a cover slip.	

(5) *Results.*—

Elastic tissue—brilliantly blue-black
Nuclei—blue-black
Collagenous fibers—red
Other tissue elements—yellow

2. **Tyson's Modification of Verhoeff's Method.**—(Poly Scientific Research and Development Corp., Deer Park, N.Y., 1975)

(1) *Fixation.*—10% Buffered formalin or Zenker's fixed tissue preferred

(2) *Section.*—5–10 μ thick

(3) *Solutions.—*

Tyson's Working Stain Solution

Mix stock solutions just before using.

A—1% Alcoholic hematoxylin	20.0 ml
B—2% Ferric chloride	25.0 ml
C—Gram's iodine	5.0 ml

And filter.

Stock Solutions

A—Alcoholic Hematoxylin Solution

Hematoxylin	1.0 g
95% Ethyl alcohol	100.0 ml

B—Ferric Chloride Solution, 2%

Ferric chloride	2.0 g
Distilled water	100.0 ml

C—Gram's Iodine Solution

Potassium iodide	2.0 g
Dissolve in distilled water	300.0 ml
Iodine	1.0 g

Dissolve potassium iodide in small amount of water, add iodine. When iodine is dissolved, add distilled water (qs) to 300 ml.

Picro-ponceau Stain (see p. 106)

Van Gieson's Stain (see p. 105)

(4) *Procedure.—*

(a)	Decerate and hydrate tissue sections to water.	
(b)	Stain in Tyson's solution Rinse in tap water and check under microscope for correct intensity of stain. Hold slides in water while examining each slide under the microscope.	20–30 min
(c)	If intensity of stain is correct, wash in slowly running tap water	5 min

(d) If overstained, differentiate in 0.5–1.0% ferric chloride solution. Rinse in tap water; examine under the microscope for correct stain intensity.

(e) When intensity of stain is correct, wash in slowly running tap water — 5 min

(f) Counterstain with picro-ponceau or Van Gieson's stain — 1–3 min

(g) Dehydrate in 95% alcohol, 2 changes

(h) Dehydrate in absolute alcohol, 2 changes

(i) Clear and mount a cover slip.

(5) *Results.*—

Elastic tissue—black
Nuclei—brown to black
Collagenous fibers—pink to red
Other tissue components—yellow

3. **Orcein Stain.**—(McManus and Mowry 1960; Humason 1972)

(1) *Fixation.*—Any general fixative but formalin or formalin-alcohol fixatives are recommended

(2) *Section.*—5–10 μ thick

(3) *Solutions.*—

Orcein Stain

Orcein, synthetic	1.0 g
70% Alcohol	100.0 ml
Hydrochloric acid, concentrated	1.0 ml

Dissolve orcein in alcohol, then add hydrochloric acid. Stain can be used immediately but staining action improves with time. Stain is stable for many months.

(4) *Procedure.*—

(a) Deparaffinize and hydrate tissue sections to water. Remove mercuric pigment if necessary.

(b) Stain in orcein stain — 30–60 min

(c)	Rinse in distilled water	a few dips
(d)	Dehydrate in 95% alcohol to remove excess stain	2 min
(e)	Differentiate in absolute alcohol for a minute or two at a time, checking under the microscope until elastic tissue is red to purple and background is almost pale brown.	

[(f), (g) and (h) are optional steps if a counterstain is desired; if counterstain is not desired, proceed to Step (i).]

(f)	Decolorize background until nearly colorless in 1% HCl in 70% alcohol Check progress of decolorization under the microscope.	2–10 min
(g)	When properly decolorized, wash in slowly running tap water	5–7 min
(h)	Counterstain with nuclear and cytoplasmic stains as desired; for example, hematoxylin and eosin, a picro-ponceau, or Van Gieson's stain, etc.; by following specific directions for each.	
(i)	Dehydrate in absolute alcohol.	
(j)	Clear and mount a cover slip.	

(5) *Results.*—

Elastic tissue—purple to red
Other tissue components—dependent on counterstains used

II. CARBOHYDRATES

Carbohydrates are aldehyde or ketone derivatives of alcohols containing more than one hydroxyl (OH) group and are constituents of many tissues. Carbohydrates can be either simple sugars or polymers of simple sugars (polysaccharides). Simple sugars are not preserved, but the polysaccharides, such as glycogen and mucopolysaccharides, can be fixed and preserved. Several staining procedures can demonstrate these components.

A. Glycogen

Glycogen is a polysaccharide of glucose found in the liver and muscle as a storage product. Glycogen can be preserved in neutral buffered formalin or in fixatives containing picric acid, such as Bouin's solution. The mechanism for preservation of glycogen by the fixatives is not exactly known, but since most carbohydrates are associated with proteins, it has been suggested that fixation promotes glycogen adsorption onto the protein in such a way as to prevent glycogen solution into the fixative.

1. **Periodic Acid-Schiff (PAS) Reaction.**—(Sheehan and Hrapchak 1973; McManus and Mowry 1960). This staining reaction is based upon a. oxidation by periodic acid (HIO_4), followed by b. staining with Schiff's reagent to form a purple-red complex.

a. **Periodic Acid Oxidation.**—The periodic acid oxidizes the adjacent free hydroxyl groups (1,2 glycol groups) of the sugar (glucose) molecule of the polysaccharide to aldehyde groups by breaking the C-C bonds at this site. The periodic acid will also oxidize 1,2 hydroxyl amino groups, 1,2 hydroxyl keto groups and 1,2 hydroxyl alkylamino groups to aldehydes. If the hydroxyl or other groups are involved in other chemical linkages the reaction will not take place.

Periodic acid is used rather than other acids, as periodic acid will not further oxidize the newly formed aldehyde groups. Aldehyde groups already present on the molecule are oxidized to carbonyl groups which do not react with Schiff's reagent. The Schiff's reagent then reacts with the aldehyde groups to form a colored complex.

b. **Staining with Schiff's Reagent.**—Schiff's reagent is not the same as colorless (leuco) basic fuchsin since the union of Schiff's reagent and the dialdehyde groups of the sugar molecule forms a new molecule whose color is purple-red and not simply the red of the oxidized basic fuchsin. This suggests a chemical change in the nature of the stain during reoxidation. The reduction of Schiff's reagent to the colorless state is obtained in the presence of high acidity and sulfur dioxide, as the basic fuchsin molecule is decolorized by the loss of the quinoid structure. When the Schiff's reagent reacts with dialdehyde, the quinoid structure is regenerated and the Schiff's reagent becomes colored.

(1) *Fixation.*—10% Neutral buffered formalin; alcoholic (absolute ethyl)-formalin (concentrated) (9:1); Bouin's solution; Zenker's solution

(2) *Section.*—5–10 μ thick

(3) *Solutions.*—

Picric Acid, 0.5% Aqueous

Periodic acid (HIO_4) crystals	0.5 g
Distilled water	100.0 ml

Schiff's Reagent

Basic fuchsin (C.I. 42500)	1.0 g
Boiling distilled water	200.0 ml
Cool, filter	
Add 1N HCl (83.5 ml of conc. HCl, sp. gr. 1.19, made to 1000.0 ml with distilled water)	20.0 ml
Sodium metabisulfite ($Na_2S_2O_5$) or anhydrous sodium bisulfite ($NaHSO_3$)	1.0g

Shake solution and keep in dark for 48 hr until solution is straw colored. Store in refrigerator (0°–4°C) in tightly stoppered bottle. Stain is good for about six months. If stain shows red color it is exhausted and should be discarded. To test reactivity of the Schiff's reagent, add a few drops of Schiff's reagent to 10 ml of 100% formalin; if solution turns red-purple immediately, it is still effective. If color development is slow and color is blue-purple, reagent has broken down.

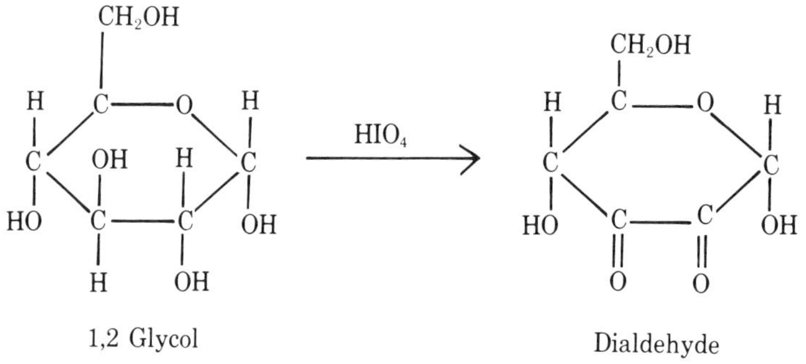

HIO_4

1,2 Glycol

Dialdehyde

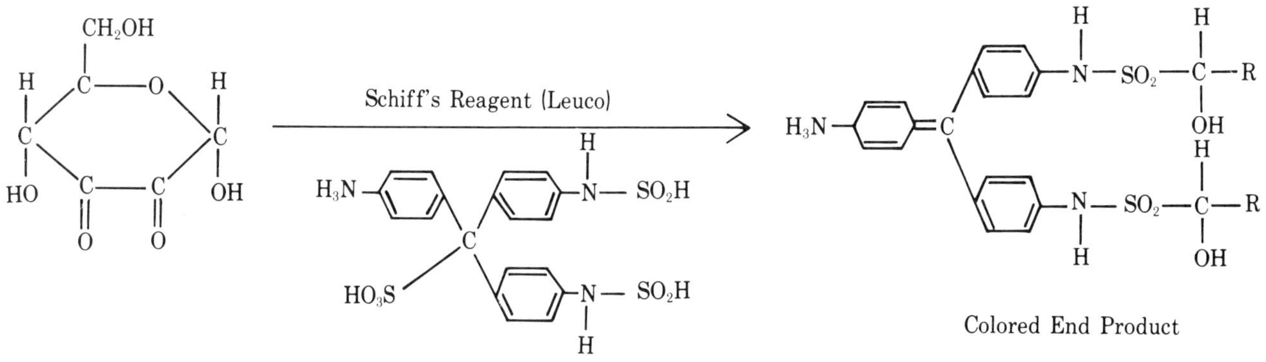

Schiff's Reagent (Leuco)

Colored End Product

(4) *Procedure.—*

(a) Decerate and hydrate tissue sections to distilled water.

(b) Oxidation of glycogen or other polysaccharide in 0.5% aqueous periodic acid — 5 min

(c) Rinse in several changes of distilled water.

(d) Stain in Schiff's reagent — 15 min

(e) Wash in slowly running tap water (Optional—rinse tissue sections in three changes of sulfurous acid, 2 min each, then wash in slowly running tap water for 10 min. The sulfurous acid rinse will reduce the overstaining by Schiff's reaction.) — 10 min

Sulfurous Acid

10% Sodium metabisulfite ($Na_2S_2O_5$)	18.0 ml
1N HCl	15.0 ml
Distilled water (qs) to	100.0 ml

Sulfurous acid should be made fresh each time it is used.

(f) Counterstain if desired—nuclei with hematoxylin and cytoplasm with plasma stains (see Exercise 7). Follow staining procedures for these stains, except time for hematoxylin staining should be no longer than 1 min as oxidation of nucleic acids increases affinity for the hematoxylin.

(g) If staining the nuclei wash in slowly running tap water — 10 min
Otherwise, omit this step. If you do not counterstain, proceed to Step (h).

(h) Dehydrate in a graded series of alcohols — a few dips in each

(i) Clear and mount a cover slip.

(5) *Results.*—

Glycogen and other reactive carbohydrates—purple to red
Nuclei and other tissue components—color of counterstains used

2. **Best's Carmine Stain.**—(Conn *et al.* 1960). This staining procedure is based on practical experience, i.e., trial and error. It is an empirical method as the staining is dependent on factors not fully explained on the basis of chemical and physical principles. The stain is unpredictable in its action, working at some times and not at other times. Less glycogen is demonstrated by this method than by the periodic acid-Schiff's reaction.

Hydrogen bonding is suggested to be the chemical reaction of the staining procedure. The carmine at pH 4.5 is converted to carminic acid, the active form of the stain. The carminic acid, negatively charged, acts as an acidic stain and is attracted to positive H on the carbohydrate molecule.

(1) *Fixation.*—10% Neutral buffered formalin; alcoholic (absolute ethyl)-formalin (concentrated) (9:1); Bouin's solution; Zenker's solution. Alcoholic fixatives are usually recommended for fixation of polysaccharides but other fixatives have been useful. The fixative should rapidly penetrate the tissue to act to destroy enzymes and to harden the tissue. Most fixatives cause glycogen to migrate to one side of the cell, but it has been suggested that the glycogen does not leave the cell since the glycogen becomes coated with protein as the protein precipitates during fixation. The cell membrane is then impermeable to the large glycogen-protein complex and glycogen remains in the cell. Loss of glycogen is thought then to be due to enzymatic hydrolysis. Furthermore, it has been observed that fixatives of alcoholic or acidic nature form coarse glycogen granules, while formalin-containing fixatives form fine glycogen droplets.

After fixation, dehydration should start in 95% or 100% alcohol rather than in lower concentrations. Attachment of paraffin sections to slides should be with 95% alcohol rather than water. Immediately after wrinkles in the tissue sections are removed, drain excess alcohol off and continue drying.

(2) *Section.*—5–10 μ thick

(3) *Solutions.*—

Carmine Solution

Stock Solution

Carmine (C.I. 75470)	2.0 g
Potassium carbonate	1.0 g
Potassium chloride	5.0 g
Distilled water	60.0 ml

Boil gently and cautiously for 5 min. Cool and filter.

Add ammonium hydroxide, conc. (28%)	20.0 ml

Refrigerate at 0°–4°C; solution lasts three months.

Working Solution

Stock carmine solution	10.0 ml
Ammonium hydroxide, conc. (28%)	15.0 ml
Methyl alcohol	15.0 ml

Solution lasts 2–3 weeks.

Differentiating Fluid

Ethyl alcohol, absolute	80.0 ml
Methyl alcohol, absolute	40.0 ml
Distilled water	100.0 ml

Celloidin or Nitrocellulose Solution, 1%[6]

Parlodion	1.0 g
Ethyl alcohol, absolute	50.0 ml
Diethyl ether, anhydrous	50.0 ml

First dissolve the Parlodion in the absolute alcohol. Stir frequently and let stand overnight. When Parlodion is completely dissolved, add the ether. Stir and frequently agitate the solution to make sure the Parlodion is completely dissolved. Store in a brown, tightly stoppered bottle to prevent deterioration and evaporation of alcohol and ether. For use this reagent must be in a liquid state.

Parlodion is highly purified cellulose tetranitrite and comes as dry strips. This highly purified substance should be used as cheaper grades are explosive.

Hematoxylin Solution (see Exercise 7)

(4) *Procedure.—*

(a)	Deparaffinize in xylene	2–3 min
(b)	Absolute alcohol	3 min
(c)	1% Celloidin solution Then dry slightly in air This step prevents diffusion of glycogen during staining.	few dips few seconds
(d)	80% Alcohol to harden the celloidin	15 sec
(e)	Rinse in distilled water	few dips
(f)	Stain in a hematoxylin solution	5 min
(g)	Wash in slowly running tap water.	
(h)	Blue nuclei in ammonia water.	
(i)	Wash in slowly running tap water.	
(j)	Stain in Best's carmine working solution	15–30 min

[6]Parlodion, Mallinckrodt Co.'s proprietary name.

(k) Differentiating fluid.
Check under the microscope to observe glycogen granules. Differentiation should take a few seconds to a few minutes. Differentiation should continue until stain no longer comes from the tissue sections.

(l)	80% Alcohol	quick dip
(m)	95% Alcohol	quick dip
(n)	100% Alcohol, 2 changes	quick dip in each
(o)	Clear in xylene-absolute alcohol (1:1)	2–3 min
(p)	Xylene, 2 changes	2–3 min in each
(q)	Mount a cover slip.	

(5) *Results.—*

Glycogen—pink to red
Nuclei—blue

B. Mucopolysaccharides

These sugars are polymerized substances composed of the disaccharide units. The disaccharide units are in turn composed of monosaccharide derivatives, hexosamine (glucosamine) and hexose if it is a neutral mucopolysaccharide, or hexosamine and uronic acid if it is an acidic mucopolysaccharide. The mucopolysaccharides can be complexed to protein, but the carbohydrate is the principal component.

The acidic mucopolysaccharides are commonly found in the mucus of the digestive system and of the salivary glands.

1. **Mayer's Mucicarmine Stain.**—(Luna 1968). In this stain the carmine "lakes" to the aluminum mordants and the aluminum binds to the acidic groups of the mucopolysaccharide. The stain demonstrates mucus from epithelial sources better than connective tissue mucus. The counterstain metanil yellow stains the connective tissue yellow, while the iron hematoxylin stains the nuclei black.

(1) *Fixation.*—Any general fixative

(2) *Section.*—5–10 μ thick

(3) *Solutions.—*

Weigert's Iron Hematoxylin

Stock Solutions A and B (see Exercise 7)

Working Solutions

Mix equal parts of Solutions A and B; prepare fresh before use.

Metanil Yellow Solution, 0.25%

Metanil yellow	0.25 g
Distilled water	100.0 ml
Glacial acetic acid	0.25 ml

Mucicarmine Stain

Stock Solution

Carmine	1.0 g
Aluminum chloride, anhydrous	0.5 g
Distilled water	2.0 ml

Mix reagents in a small test tube. Heat for 2 min. Liquid becomes black and syrupy.

Dilute with 50% alcohol	100.0 ml

Let stand for 24 hr and filter. Solution is stable for several months.

Working Solution

Stock mucicarmine solution	1 part
Tap water	4 parts

Diluted mucicarmine deteriorates within a few hours; prepare solution just before using.

(4) *Procedure.—*

(a)	Deparaffinize and hydrate slides to distilled water. Remove mercuric pigment if necessary.	
(b)	Stain in working solution of Weigert's iron hematoxylin	7–10 min
(c)	Wash in slowly running tap water	5–10 min
(d)	Stain in working solution of mucicarmine Check under the microscope after 30 min, and after that at 10 min intervals, to obtain correct depth of color for the	30–60 min

mucus. Rinse briefly in distilled water before checking. Hold slides in water while checking.

(e) When correct stain intensity is reached, rinse quickly in distilled water — few dips

(f) Counterstain in metanil yellow — 1 min

(g) Rinse quickly in distilled water — few dips

(h) Rinse quickly in 95% alcohol — few dips

(i) Dehydrate in absolute alcohol, 2 changes

(j) Clear with xylene, 2–3 changes

(k) Mount a cover slip.

(5) *Results.*—

Mucus—deep rose to red
Nuclei—black
Other tissue components—yellow

2. **Periodic Acid-Schiff's Reaction** **(See p. 113)**

III. SPECIAL CELLS

1. **Berg's Method for Spermatozoa.**—(Sheehan and Hrapchak 1973; Luna 1968). In this staining method, new fuchsin substitutes for basic fuchsin for acid-fast material. The spermatozoa, which are acid-fast, are stained bright red, while other tissue components are blue to purple after counterstaining with methylene blue. Another modification of this technique from the usual acid-fast staining procedure is to use a weaker differentiating reagent which removes less stain than the concentrated hydrochloric acid-alcohol solutions.

(1) *Fixation.*—10% Neutral buffered formalin suggested

(2) *Section.*—5–10 μ thick

(3) *Solutions.*—

Putt's Carbol-fuchsin Solution

New fuchsin	1.0 g
Phenol	5.0 g
Ethyl alcohol, 100%	10.0 ml
Distilled water	84.0 ml

Lithium Carbonate, 1%, Saturated

Lithium carbonate	1.0 g
Distilled water	100.0 g

Glacial Acetic Acid (5%)-Alcohol Differentiating Solution

Glacial acetic acid	5.0 ml
Ethyl alcohol, 100%	95.0 ml

Methylene Blue, 0.5%

Methylene blue	0.5 g
Ethyl alcohol, 100%	100.0 ml

(4) *Procedure.*—Agitate slides throughout the staining procedure to prevent uneven staining of the tissue sections.

(a)	Deparaffinize and hydrate tissue sections to distilled water.	
(b)	Stain in carbol-fuchsin solution	3 min
(c)	Place slides directly into the lithium carbonate solution to mordant the stain	3 min
(d)	Differentiate spermatozoa by placing tissue sections into acetic acid-alcohol differentiating solution to decolorize the other tissue components	5 min
(e)	Absolute ethyl alcohol, 2 changes	1 min in each
(f)	Counterstain with methylene blue	30–60 sec
(g)	Absolute ethyl alcohol, 2 changes	rapid rinse in each
(h)	Clear in xylene, 2–3 changes	few dips in each
(i)	Mount a cover slip.	

(5) *Results.*—

Spermatozoa—bright red
Erythrocytes—pale pink
Other tissue components—blue to purple

(See also eosin/nigrosin technique.)

IV. NUCLEIC ACIDS

A staining procedure used to distinguish the degree of polymerization of nucleic acids and so differentiate DNA from RNA is methyl green-pyronin Y. A suggested chemical explanation for the differential staining of these two nucleic acids by these stains may be due to the degree of polymerization possessed by the two molecules. The highly polymerized DNA is stained by methyl green, while the pyronin stains the smaller sized RNA. Depolymerization of DNA promotes binding of pyronin and loss of the DNA's ability to bind methyl green. The methyl green is bound to two phosphate groups of the DNA by its two amino groups.

As the pyronin is not necessarily specific for RNA, ribonuclease enzymatic digestion should be run in parallel sections as control slides in order to specifically localize the RNA. Those tissue components stained initially red with pyronin but no longer red after enzymatic digestion are considered to contain RNA. Those components which are still stained red after the enzymatic digestion are not considered to contain RNA. If the enzymatic digestion step is to be employed, do not fix the tissue in fixatives containing heavy metal ions, as the ions will complex with the RNA and render it resistant to the enzymatic digestion.

For research purposes, commercially obtained methyl green should be purified before use by extracting the impurity, methyl violet (crystal), with chloroform several times. The methyl violet will dissolve in the chloroform whereas the methyl green remains in the aqueous phase. For routine use of the stain, purification is not necessary. The unextracted stain gives a bluer coloration to the tissue components as it stains.

1. **Methyl Green-pyronin Y Stain.**—(Sheehan and Hrapchak 1973; Humason 1972; Davenport 1960; Europa 1976)

(1) *Fixation.*—Any alcoholic fixative, such as Carnoy's fixative or 10% neutral buffered formalin

(2) *Section.*—5–10 μ thick

(3) *Solutions.*—

Methyl Green-pyronin Y Solution

Methyl green (purified, see below)	0.5 g
Pyronin Y	0.2 g
Sodium acetate-acetic acid buffer, 0.2 M (pH 4.1)	100.0 ml

Dissolve the methyl green in the sodium acetate buffer, then add the pyronin and mix well with a magnetic stirrer. Solution is stable for months in refrigerator and is reusable.

Purification of Methyl Green Solution

Dissolve 0.5 g methyl green in 100.0 ml of 0.2 M sodium acetate-acetic acid buffer (pH 4.1). Extract the methyl violet from the methyl green in a separatory funnel with at least four changes of chloroform or until no more violet color is removed from the solution. Evaporate the residual chloroform from the methyl green solution by exposure to the air; usually overnight is sufficient.

Sodium Acetate-acetic Acid Buffer, 0.2 M, pH 4.1

0.2 M Acetic Acid

Glacial acetic acid	5.7 ml
Distilled water	500.0 ml

0.2 M Sodium Acetate

Sodium acetate ($CH_3COONa \cdot 3H_2O$)	13.6 g
Distilled water	500.0 ml

Mix 3 parts acetic acid to 1 part sodium acetate to give a pH about 4.1.

(4) *Procedure.—*

(a)	Deparaffinize and hydrate tissue sections to distilled water.	
(b)	Rinse in 3 changes of distilled water	a few dips in each
(c)	Place in 0.2 M sodium acetate-acetic acid buffer, pH 4.1	15 min
(d)	Stain—without rinsing, place tissue sections directly into the methyl green-pyronin Y staining solution	1–2 hr
(e)	Drain slides and rinse in ice-cold (0°C) distilled water	very quickly
(f)	Blot with bibulous paper.	
(g)	Dehydrate in acetone . . . and	2 quick dips
(h)	Acetone-xylene mixture (1:1)	1 min
(i)	Clear in xylene, 2 changes	3 min in each
(j)	Mount a cover slip.	

(5) *Results.—*

Chromatin material (DNA)—green to blue-green
Nucleolar and cytoplasmic RNA—red to red-purple

V. EXFOLIATIVE CYTOLOGY

Exfoliative cytology is the study of cells that have been shed (desquamated) or removed from an epithelial surface from such organs as the vagina, cervix and uterus of the female reproductive system, prostate gland of the male reproductive system, urinary, digestive and respiratory systems, and from sputum, pleural and peritoneal fluids. The cells can be collected from the epithelial surfaces by lightly scraping the surface or by swabbing, aspirating or washing the surface. The cells are then smeared on a glass slide, fixed, stained, and, after coverslipping, are ready for microscopic examination.

The primary purpose of examination of these cells is a rapid screening procedure for the diagnosis of cancer, as it has been demonstrated that there is an association between the presence of shed abnormal cells in the vaginal smear and the presence of pre-malignant conditions of the cervix. The procedure has other purposes also. It is used to determine the proper physiological levels of sex hormones in the female as the sex hormones change the character of the vaginal epithelium during phases of the menstrual cycle. However, these vaginal cytology tests have lost much of their value with the use of hormonal contraceptives. Furthermore, the procedure can indicate presence of infection in the reproductive tract and determine the genetic sex of the individual (Barr bodies).

Normally, epithelial cells are desquamated and their morphology has been described (see following). During malignant conditions or during infection, the desquamation becomes exaggerated and the epithelial cells become larger in size. Thus by studying the alteration in cytology of the epithelial cells and the desquamation patterns, diagnosis of various pathologic conditions may be made. Although smears provide for a rapid diagnosis, a positive cytological examination is not always conclusive and a biopsy should be taken for further confirmation of the diagnosis.

A. Normal Vaginal Epithelial Cell Cytology

The vaginal mucosa, which consists of three layers, is composed of nonkeratinized stratified squamous epithelium. From this mucosal layer most of the cells are shed and are seen in the smear. The vaginal epithelium undergoes cyclic changes to correspond to the changes in the hormonal levels of estrogen and progesterone during the menstrual cycle. The appearance of the superficial and intermediate cells at the proper times in the menstrual cycle indicates sufficient levels of sex hormones.

The superficial layer is made up of several layers of mature squamous cells. In the smear, these cells are large (40–50 μ diameter), flat, polygonal shaped cells. The cytoplasm of a majority of these cells is eosinophilic or acidophilic as it stains red-pink; granules are sometimes present in the cytoplasm. Other superficial cells, which come from lower down in the superficial layer, have basophilic cytoplasm which stains green-blue. The nuclei of the cells of the superficial layer are dark staining, shrunken, condensed (pyknotic) structures. A high percentage of the superficial cells of the smear demonstrating the cytology described is a good indication of adequate estrogen activity. If estrogen levels are not high enough, maturity is not achieved and intermediate cells form the majority of the smear.

The middle layer is the intermediate layer consisting of several layers of less mature cells. These cells are about the same size or slightly smaller than superficial cells. Their shape is flat and polygonal but some cells are boat shaped or "navicular" shaped. The cytoplasm of most of these cells is basophilic as it stains green to blue. Some intermediate cells demonstrate eosinophilia. The cytoplasm of these cells has been demonstrated to contain glycogen. The nuclei of the intermediate cells are relatively larger than superficial cell nuclei and are spherical to oval in shape. These nuclei are vesicular, i.e., they are light blue stained with observable chromatin granules.

The innermost layer of the mucosa is composed of two sublayers, the parabasal and basal layers. The parabasal cells are normally not as common in the smear as cells of the upper two layers and increase in the smear during many pathological conditions. These cells are smaller than the intermediate cells but more spherical in shape, with basophilic cytoplasm. If cells are clumped together, it means that the cells were scraped off during removal of the smear specimen. The parabasal cell nuclei are large and vesicular. The basal cell layer, resting on the basement membrane, is rarely seen in the smear unless pathological conditions have damaged the more superficial cell layers. These cells are spherical to oval in shape and have a small amount of basophilic cytoplasm with a relatively large vesicular nucleus.

B. Cytological Criteria of Malignancy

Malignant cells show nuclear features which are not seen in normal cells, such as changes in nuclear/cytoplasmic ratio and nuclear morphology. Nuclei of malignant cells are large and occupy most of the cell; cytoplasm of these cells is scanty and hardly visible; thus an increase in nuclear/cytoplasmic ratio. The nuclei are irregular in size, shape and staining intensity. Most nuclei stain darkly (hyperchromatic), with clumping of chromatin and prominent nucleoli. The boundary of the nuclei is sharply outlined and appears crinkled. The nuclei also show abnormal mitoses. In general, the cells of the smear lack a uniformity of appearance among the cells.

1. **Papanicolaou, "Pap," Smear Method.**—(Humason 1972; Emmel and Cowdry 1964; Papanicolaou 1942). The most widely used method today in exfoliative cytology is the Papanicolaou or "Pap" smear method. In the classroom situation to demonstrate the staining procedure scrapings are made from the inside of the cheek of the buccal cavity (oral mucosa) with a flat end of a toothpick first dipped in 70% alcohol. The shed cells are immediately and uniformly smeared on a glass slide and fixed as the exfoliated cells rapidly degenerate. Furthermore, the smear must not become dry during any time of the procedure as the cells' morphology becomes distorted.

   a. **Solutions.**—

   Fixative

   95% Ethyl alcohol-ethyl ether 1:1 mixture

Staining Solutions

These stains can be prepared or purchased already prepared from various scientific supply companies.

Harris' Hematoxylin Without Acetic Acid
(See Exercise 7 for its preparation.)

Filter and store in dark bottle. Filter working solution regularly and replenish often by adding small amount of fresh stock to maintain uniform staining results. Working solution is replaced every 1 to 3 weeks depending on quantity of slides being stained.

Orange G 6 (OG-6)

Orange G 6, 0.5% (0.5 g/100 ml 95% alcohol or S-29)	100.0	ml
Phosphotungstic acid	0.015	g

Eosin-azure 36 (EA-36)

Light green SF (yellowish), 0.5% (0.5 g/ml 95% alcohol or S-29)	45.0	ml
Bismarck brown (C.I. 21000), 0.5% (0.5 g/100 ml 95% alcohol or S-29)	10.0	ml
Eosin Y, 0.5% (0.5 g/100 ml 95% alcohol or S-29)	45.0	ml
Phosphotungstic acid	0.2	g
Lithium carbonate, saturated aqueous solution (1.25%)	1	drop

(1) *Fixation.—*

(a) After scraping the epithelium, smear cells on a glass slide and immediately immerse smear into 1:1 mixture of 95% alcohol-ethyl ether — 15 min

(2) *Hydration.—*

(b) 95% Ethyl alcohol — 2 min

(c) 80% Ethyl alcohol — 2 min

(d) 70% Ethyl alcohol — 2 min

(e) Tap water — 2 min

(f) Distilled water — 2 min

(3) *Staining and Counterstaining.—*These steps may be carried out manu-

ally or by automatic staining procedures. The procedure below lists the staining steps used in the automated Papanicolaou method in the Mono-Autotechnicon.

(a)	Distilled water	2 min
(b)	Harris' hematoxylin	2–4 min
(c)	Distilled water	5 min
(d)	Dehydrant (S-29)	2.5 min
(e)	Dehydrant	2.5 min
(f)	Orange G 6 (OG-6)	2.5 min
(g)	Dehydrant	2.5 min
(h)	Dehydrant	2.5 min
(i)	Eosin-azure 36 (EA-36)	2.5 min

(4) *Hydration.—*

(j)	Dehydrant	2.5 min
(k)	Dehydrant	2.5 min

(5) *Clearing.—*

(l)	Xylene	2–5 min
(m)	Xylene	2 min

(6) *Mounting.—*

(n) Mount a cover slip.

(7) *Results.—*

Nuclei—blue
Acidophilic cells—red to orange
Basophilic cells—green to blue-green

Harris' hematoxylin or any other hematoxylin can be used. Do not overstain nuclei as chromatin and nucleoli will be obscured. Darkly stained nuclei give a false impression of hyperchromasia. If nuclei are understained the malignant character of the nuclei may be overlooked. The first counterstain, orange G 6, colors the cytoplasm orange if cells are keratinized. The second counterstain is a polychrome mixture of eosin, light green and bismarck brown. Cells having acidophilic cytoplasm, such as the keratinized superficial squamous cells, show an affinity for eosin and take on shades of pink to yellow. Those with basophilic cytoplasm are stained greenish-blue by the light green. Various factors, such as drying, pH, thickness of the smear, etc., may alter cytoplasmic staining reactions.

QUESTIONS

Cover answers with a piece of paper. Answers appear at end of questions.

(1) The Mallory and Masson staining procedures are excellent for demonstrating
(a) DNA
(b) RNA
(c) Collagenous fibers
(d) Carbohydrates

(2) Verhoeff's stain is good for demonstrating
(a) Orcein
(b) Muscle
(c) RNA and DNA
(d) Elastic tissue

(3) The periodic acid-Schiff reaction (PAS) is a good stain for
(a) Glucose
(b) Periodic acid
(c) Glycogen
(d) Quinoids

(4) Mayer's mucicarmine stain is a good stain for demonstrating
(a) Glycogen
(b) Luna bodies
(c) Mucopolysaccharides
(d) Laking of metanil yellow

(5) Berg's staining method is used to demonstrate
(a) Sperm cells
(b) Vaginal epithelium
(c) Cervical epithelium
(d) Muscle tissue

(6) The methyl green-pyronin Y stain is used to demonstrate
(a) Carnoy bodies
(b) Nissl bodies
(c) Ribonuclease
(d) DNA and RNA

(7) The correct spelling of the originator of the concept of exfoliative cytology is
(a) Papanicolaou
(b) Papinacoloau
(c) Papenheimer
(d) Heidenheimer

(8) Nuclei which are dark staining, shrunken and condensed are said to be
(a) Squamous
(b) Pyknotic
(c) Navicular
(d) Parabasal

(9) The stain orange G 6 is used
(a) To demonstrate hyperchromatic nuclei
(b) As part of the exfoliative cytology staining procedure
(c) For staining spermatozoa
(d) For demonstrating DNA

(10) A staining procedure which is based on trial and error and not fully explained on the basis of chemical and physical factors is
(a) Iron hematoxylin
(b) Heidenhain's hematoxylin
(c) Best's carmine stain
(d) Schiff's stain

Answers

(1) c	(6) d
(2) d	(7) a
(3) c	(8) b
(4) c	(9) b
(5) a	(10) c

11

Cryostat Sectioning Technique and Staining

Cryostat or frozen section technique is used in the field of histochemistry to localize and identify inorganic and organic chemicals, especially enzymes, at specific sites at which they occur in cells and tissues. Some histochemical procedures that employ frozen sections are enzymatic studies, autoradiography and fluorescent antibody techniques. Frozen sections are also used on biopsy tissue to see if further surgery is required.

Limitations of the frozen section technique due to the inability to maintain tissue block, knife and tissue sections at the same temperature when using the clinical freezing microtome (see Fig. 11.1) have been overcome by the use of the cryostat (see Fig. 11.2). The cryostat rapidly obtains thin sections of fresh tissue, which are ready for histochemical studies, with a quality equal to or better than paraffin technique processed tissue sections.

This exercise will discuss how frozen sections are made on the cryostat and will not describe any histochemical procedures. The technologist should consult the literature for specific histochemical procedures.

I. CRYOSTAT SECTIONING

Cryostat sectioning requires only a minimum of training once the technologist masters the operation of the standard rotary microtome.

A. Cryostat

The cryostat is an insulated, thermostatically controlled refrigerated cabinet in which a rust-proof rotary microtome is housed. The more recently designed cryostats are open top models, with the rotary handle of the microtome located external to the cabinet. The open top and exterior position of the microtome's handle allows for ease of operation and manipulation of sections during the cutting.

B. Operating Temperature

The temperature of the cabinet is kept between −15°C and −20°C. Because of

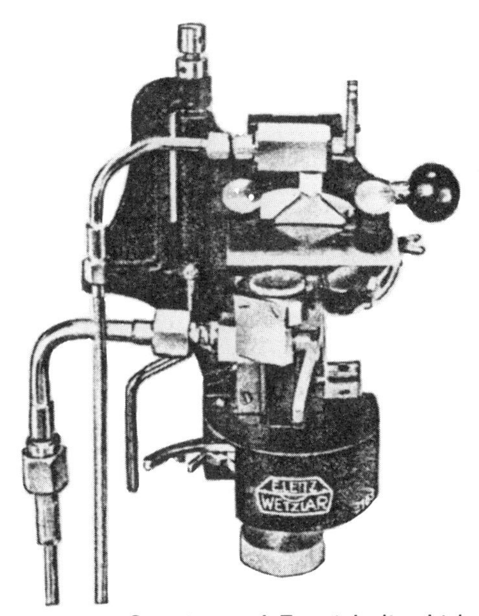

Courtesy of Ernst Leitz, Ltd.

FIG. 11.1. CLINICAL FREEZING MICROTOME WITH LOW TEMPERATURE KNIFE COOLING

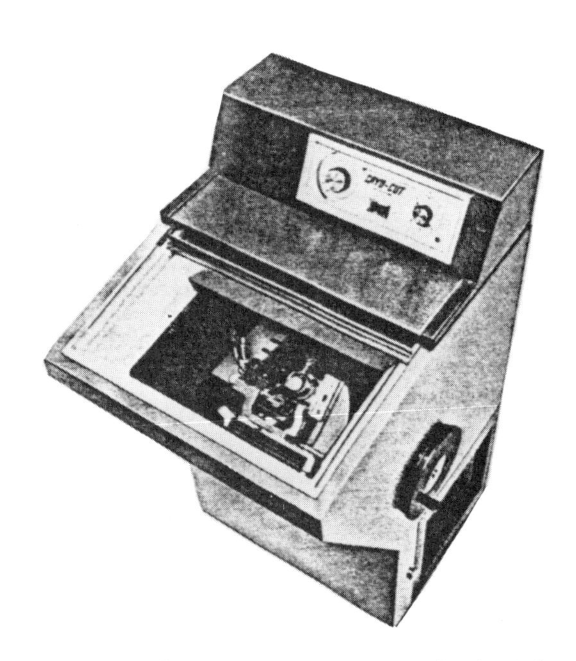

Courtesy of American Optical Corp.

FIG. 11.2. CRYOSTAT (AMERICAN OPTICAL CORP., BUFFALO, N.Y.)

difference in composition of tissues, the temperature of sectioning varies, but the optimal working temperature is between −18°C and −20°C for most tissues.

Fibrous tissue cuts easier than soft tissues. If tissues are too cold they fragment on sectioning or if too warm they collapse at the knife edge.

The microtome is mounted deep within the cryostat cabinet at an angle of 45° and with the top open during sectioning the temperature where sectioning is occurring is not sufficiently raised to interfere with the sectioning. The efficiency of sectioning in the cryostat depends only on similar temperature of the tissue block and knife.

C. Preparation of Tissue for Sectioning (Quenching)

1. **Size of Tissue Samples.**—Fresh tissues, no more than 2–4 mm thick, are removed from the organism and are quickly frozen at very low temperatures. Small pieces of tissue and rapid freezing are essential to prevent post-mortem changes and diffusion of substances within the tissues. The extremely low temperatures and rapid freezing are important to prevent formation of large ice crystals, which would disrupt cell structure. The use of small samples of tissue overcomes these problems.

2. **Quenching Procedures.**—Rapid freezing of the tissue can be accomplished by several methods.

   a. **Isopentane and Liquid Nitrogen.**—In this method the tissue sample is placed on aluminum foil and the tissue is plunged into isopentane or Freon 12 which is cooled to −160°C to −180°C with liquid nitrogen. The frozen tissue is then attached to precooled metal tissue holders with a few drops of water, saline or blood serum which almost instantly holds the tissue in place.

   b. **Cryostat's "Heat Sink" and Heat Extractor.**—Another method for freezing tissue is to use within the cryostat cabinet the "heat sink" and also attached heat extractor (see Fig. 11.3). The metal tissue holders, to which the tissue is attached, are held in the slots of the heat sink to be cooled. The heat sink and heat extractor have a high thermal conductivity and draw heat off from the tissue sample rapidly because of the relatively large mass of cold metal. The fresh tissue can be attached to the precooled metal tissue holders seated on the heat sink as already described, the heat extractor is then placed over the tissue, and in 30–60 seconds, the tissue is frozen.

      Alternatively, optimal cutting temperature (O.C.T.) compound, composed of synthetic glycols and resins, can be used to attach and support the tissue on the tissue holder. The O.C.T. at room temperature is viscous, and when frozen is a solid opaque material. The O.C.T. when frozen acts to attach tissue to the tissue holders and supports tissue during cutting. In this regard, the O.C.T. acts as an "embedding medium." If O.C.T. is used, place tissue in the O.C.T. on the precooled tissue holder; the heat extractor is rested on the tissue and the tissue is frozen.

   c. **Carbon Dioxide Quick Freezing Chamber.**—An even faster freezing procedure requiring about 10 seconds comprises the use of compressed carbon dioxide gas or "dry ice" using a quick freezing chamber attached to the carbon dioxide gas cylinder (see Fig. 11.4). Place a precooled metal tissue holder into the quick freeze chamber; squeeze some O.C.T. from its dispenser onto the face of the tissue holder to form a base. Put the tissue in this O.C.T., and add some more to cover the tissue. Then with several (2–5) short bursts of carbon dioxide gas from the cylinder freeze the tissue. Hold the plastic holder down lightly over the tissue to prevent it from being blown away.

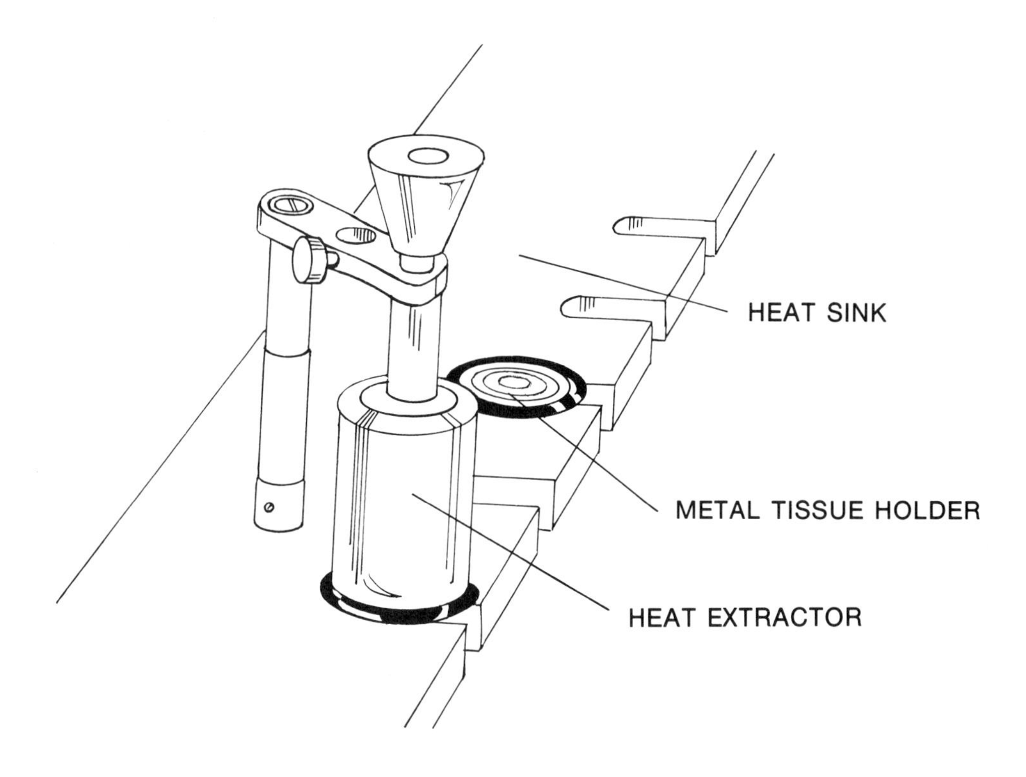

FIG. 11.3. "HEAT SINK" AND HEAT EXTRACTOR

When tissue is frozen, place it in the heat sink of the cryostat with the heat extractor over the tissue for a short time.

Once the tissue is frozen by any method, tissue can be stored on the heat sink, but be sure to cover the tissue with aluminum or plastic wrap to prevent freezer burn.

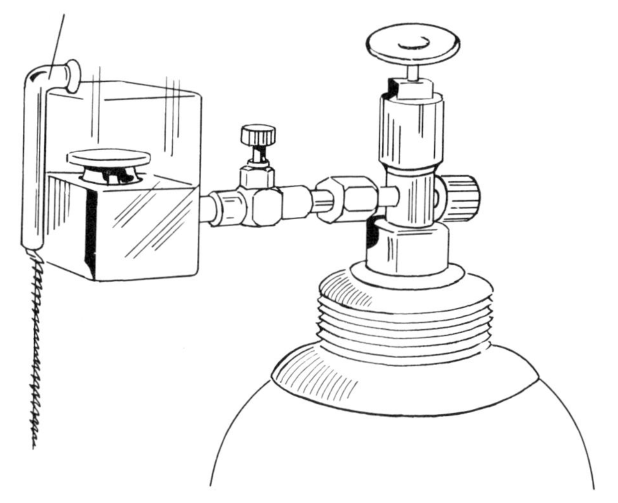

FIG. 11.4. CARBON DIOXIDE QUICK FREEZING CHAMBER AND CYLINDER

D. Sectioning of the Frozen Tissue

After the tissue is frozen, the tissue holder is clamped to the microtome's chuck and the tissue is adjusted as for paraffin sectioning technique.

A sharp microtome knife is inserted into the knife holder and must be precooled several hours before sectioning. The angle or tilt of the knife between the knife edge and face of the block is about 20°–30°.

The section thickness is adjusted so the microtome cuts the tissue sections between 4 and 6 μ thick.

Release the brake on the rotary handle, rotate handle so tissue is brought in line with the knife edge, then using the tissue feed carrier knob or fast feed control, rapidly advance the tissue until it almost contacts the knife edge. Then with the rotary handle of the microtome, trim off excess O.C.T. until the face of the tissue is exposed. Clear knife of debris with a cold camel's hair brush and begin to section.

To make a section, place the plastic antiroll plate or guide over the knife face. The antiroll plate prevents the section from curling as the cold temperature makes the sections roll up as they are cut from the tissue block. The edge of the antiroll plate must be adjusted to be parallel and even to the knife edge, since it must guide the cut section under itself and the face of the knife. Two raised ridges on the antiroll plate allow for a slight separation through which the section slides.

Alternatively, the section can be cut and guided onto the knife face with a camel's hair brush as the tissue moves down over the knife.

The cryostat's microtome (see Fig. 11.5) cuts individual sections and does not form ribbons. Sectioning is best accomplished by turning the microtome's rotary handle with a slow, firm, deliberate stroke and not with the continuous rhythmic motion of paraffin sectioning. The wheel of the handle is grasped with the entire hand, with the knob of the handle seated between the thumb and second finger.

If all adjustments are properly made and surfaces are absolutely clean, the section will glide smoothly between the knife face and the antiroll plate and lie flat beneath the antiroll plate. An acceptable tissue section is one that lies flat on either the knife face or antiroll plate.

There are several theoretical factors that determine if a flat section will be made.

1. Clean surface is critical as a film of condensed vapor produces enough resistance to cause the section to crease as it glides between the antiroll plate and knife face.

2. On cutting a section, the heat generated by the cutting edge of the knife melts a very thin layer of the tissue (650 Å) on the cut section as well as on the face of the tissue block. The melted tissue refreezes as it passes down the knife surfaces. The rate of refreezing of this layer of tissue just past the knife edge is critical for a flat section to be made. The lower the knife temperature the thinner the melted layer of tissue, as the heat of cutting is conducted directly into the knife.

3. Rigidity of the section is also critical, i.e., how cold the section is, and its thickness will determine whether the section can withstand the frictional resistance as it slides over the knife face.

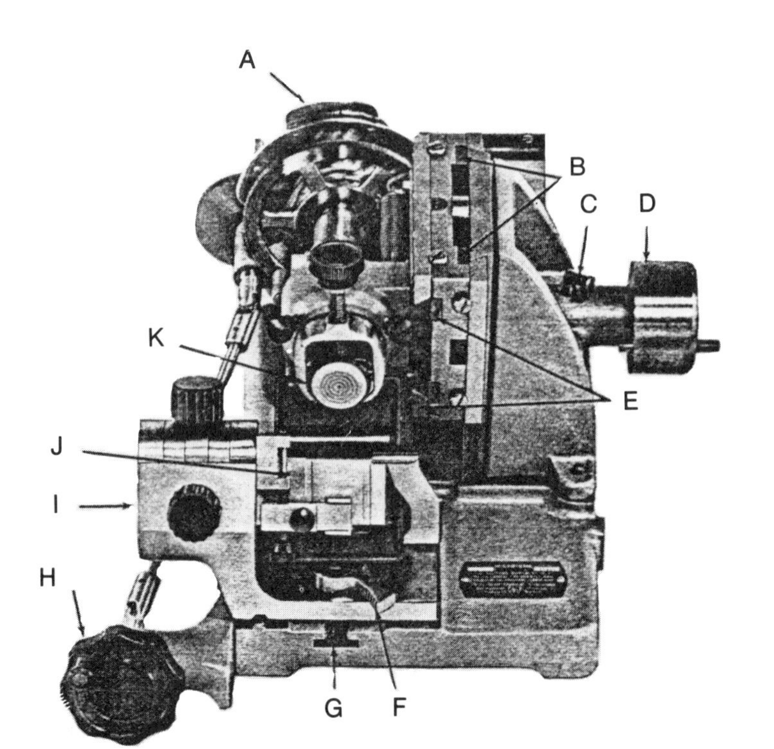

Courtesy of American Optical Corp.

FIG. 11.5. A. O. CRYO-CUT MICROTOME (AMERICAN OPTICAL CORP., BUFFALO, N.Y.)

E. Handling of the Tissue Sections

Once the tissue section has been cut, the antiroll plate is lifted away and the section should lie flat on the knife face. The section can now be attached to a glass slide or cover slip. It is easy to pick up the cold tissue section on a warm cover slip. The cover slip is held by a suction device (see Fig. 11.6). The warm cover slip is placed over the tissue section and the tissue section is attracted to the warm cover slip. The O.C.T. melts and the tissue section remains attached to the cover slip. The suction is broken between the suction device and the cover slip by placing a fingernail under the cover slip. The cover slip is then picked up by a flat jawed cover slip forceps (see Fig. 11.7B) and the tissue is carried through a staining procedure or some histochemical procedure (see following Section II).

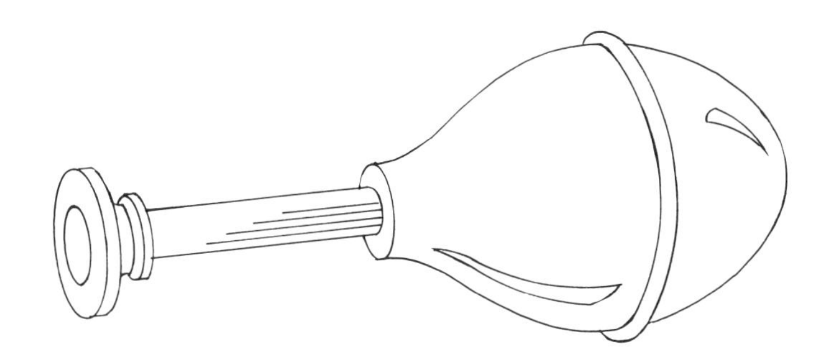

FIG. 11.6. SUCTION DEVICE FOR HOLDING COVER SLIP

Before another tissue section is cut on the cryostat, the frost mark where the section rested is brushed away with a cold camel's hair brush. The antiroll plate is repositioned and another section is cut.

II. STAINING

There are many staining and histochemical procedures which can be performed on fresh unfixed frozen tissue. Only one will be described so that the finished product of cryostat sectioning technique can be observed.

Staining is carried out in small Columbia staining jars (see Fig. 11.7A) which hold the 22 mm^2 cover slips. The cover slips are moved from one jar to the next with the cover slip forceps. Always keep tissue facing you so you know which side of the cover slip the section is on, as the section is so thin that it is hard to recognize.

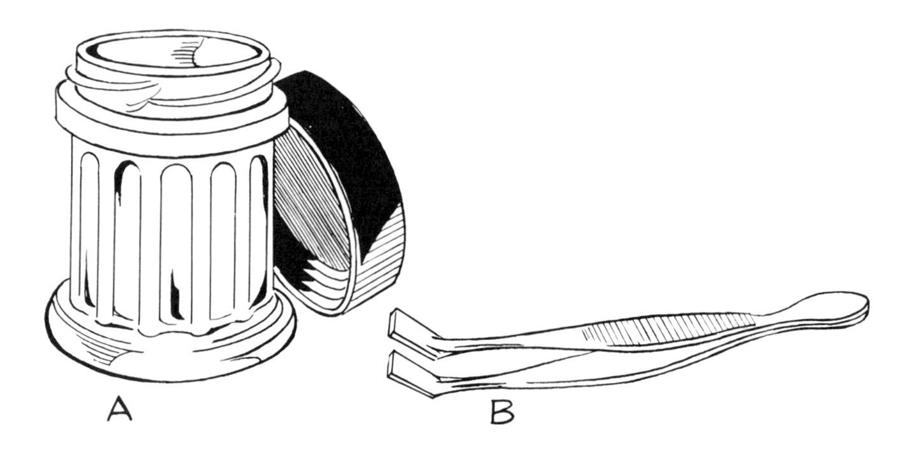

FIG. 11.7. A—COLUMBIA STAINING JAR
B—COVER SLIP FORCEPS

A. Hematoxylin-eosin Staining for Cryostat Sections

1. Post-fixation—Optional Step
   Acetic-alcohol (glacial acetic acid:100% ethyl alcohol, 1:3) — 15 sec
   This fixative penetrates rapidly, gives good nuclear fixation and adequate preservation of cytoplasmic structures.

2. Distilled water — 15 sec

3. Staining in hematoxylin (Delafield's or Harris') — 30–45 sec

4. Distilled water

5. Tap water; rinse until tissue section is blue.

6. Mordant with saturated mercuric chloride — 30 sec

7.	Counterstain with 0.5% eosin Y in 95% ethyl alcohol	30 sec
8.	Dehydrate and wash off excess eosin in 95% ethyl alcohol	quick dip
9.	Dehydrate and further blue the nuclei in 100% ethyl alcohol saturated with lithium carbonate, 2 changes	15 sec in each
10.	Clear with xylene	30 sec or until sections are clear
11.	Mounting: Streak cover slip with permount and place cover slip with the stained tissue section down on a glass slide.	

Rapid staining can also be carried out in toluidine blue or methylene blue.

III. ADVANTAGES AND DISADVANTAGES OF CRYOSTAT TECHNIQUE

A. Advantages of Cryostat Technique

1. In histochemistry where enzymes are studied, the enzymes are preserved by this technique, as enzymes are destroyed above 56°C.
2. There is no shrinkage of the tissue as tissue is not processed through dehydrant or clearing agents.
3. Lipids can be demonstrated as they are not dissolved out as in paraffin technique.
4. Rapid diagnosis can be made during course of an operation where malignancy is suspected. With the patient under anesthesia, the operation is suspended until the diagnosis is made on the biopsy (particularly significant in breast surgery).

B. Disadvantages of Cryostat Technique

1. Can not prepare a ribbon of tissue sections as cryostat temperature is too cold; heat of sectioning does not sufficiently melt the O.C.T. to have the sections stick together.
2. Difficult to prepare serial sections of the tissue as sometimes the cryostat microtome does not always make a section on turning the rotary handle, or the sections do not glide smoothly under the antiroll plate.
3. Lack of infiltration and embedding leads to distortion of structural details upon sectioning the tissue.
4. Staining of unfixed tissue does not give the same clarity of detail as fixed tissue.

QUESTIONS

Cover answers with a piece of paper. Answers appear at end of questions.

(1) In what area of histology is cryostat section technique useful?
- (a) "Pap" smear
- (b) Histochemistry
- (c) Vacuum infiltration
- (d) All of the above

(2) Histochemistry allows one to
- (a) Identify specific cells at the sites they occupy in the tissue
- (b) Section frozen tissues
- (c) Localize cellular organelles only
- (d) Localize and identify inorganic and organic chemicals in cells and tissues

(3) What are the two main components of the cryostat?
- (a) Rotary microtome; refrigerated cabinet
- (b) Refrigerated cabinet; clinical freezing microtome
- (c) "Heat sink"; rotary microtome
- (d) Carbon dioxide quick freeze chamber; heat extractor

(4) What is the optimal temperature for sectioning in the cryostat?
- (a) +15°C to +20°C
- (b) 0°C to −5°C
- (c) −18°C to −20°C
- (d) −15°C to −10°C

(5) Choose from the list below two ways tissue samples may be quick frozen
- (a) Isopentane in liquid nitrogen; compressed carbon dioxide
- (b) O.C.T.; isopentane
- (c) "Heat sink"; liquid nitrogen
- (d) None of the above

(6) In what capacity does optimal cutting temperature (O.C.T.) compound act in the cryostat technique?
- (a) To freeze the tissue
- (b) As an infiltration medium
- (c) To attach the sections to a cover slip
- (d) As an "embedding" medium

(7) What is the purpose of the plastic guide over the knife face?
- (a) Prevents sections of the tissue block
- (b) Prevents tissue section from curling as it is cut from face of the tissue block
- (c) Helps to attach sections to cover slips
- (d) There is no purpose to this guide

(8) On sectioning, the cryostat makes
- (a) Individual sections
- (b) A ribbon of sections
- (c) Sections all the time
- (d) None of the above

(9) Why is a clean cryostat microtome's knife surface critical?
   (a) A film of condensed vapor produces heat to melt the sections on the knife's surface
   (b) The thickness of the section causes the heat of sectioning to crease the section as it contacts the knife surface
   (c) A film of condensed vapor produces resistance to crease the sections
   (d) Heat of sectioning causes sections to stick together

(10) In what direction should the tissue sections be held during their staining?
   (a) Upside down
   (b) Facing you
   (c) Inverted
   (d) Away from you

Answers

(1) b
(2) d
(3) a
(4) c
(5) a
(6) d
(7) b
(8) a
(9) c
(10) b

REFERENCES

ANDREWS, G.S. 1971. Exfoliative Cytology. C.C. Thomas, Springfield, Ill.

BAKER, J.R. 1966. Cytological Technique, 5th Edition. Methuen and Co., London.

COHN, A. 1975. Handbook of Microscopic Anatomy for the Health Sciences. C.V. Mosby, St. Louis.

CONN, H. J. 1953. Biological Stains, 6th Edition. Biotech Publications, Geneva, N.Y.

CONN, H.J., DARROW, M.A., and EMMEL, V.M. 1960. Staining Procedures, 2nd Edition. Williams and Wilkins, Baltimore.

CULLING, C.F.A. 1974. Handbook of Histopathological and Histochemical Techniques, 3rd Edition. Butterworths, Woburn, Mass.

DAVENPORT, H.A. 1960. Histological and Histochemical Technics. W.B. Saunders, Philadelphia.

DRURY, R.A.B., and WALLINGTON, E.A. 1967. Carleton's Histological Technique, 4th Edition. Oxford University Press, New York.

EMMEL, V.M., and COWDRY, E.V. 1964. Laboratory Technique in Biology and Medicine, 4th Edition. Williams and Wilkins, Baltimore.

EUROPA, D. 1976. Bellevue modified methyl green-pyronin stain. Personal communication. Bellevue Hospital, New York.

GALIGHER, A.E., and KOZLOFF, E.N. 1971. Essentials of Practical Microtechnique, 2nd Edition. Lea and Febiger, Philadelphia.

HUMASON, G.L. 1972. Animal Tissue Techniques, 3rd Edition. W.H. Freeman, San Francisco.

LUNA, L.G. 1968. Manual of Histologic Staining Methods of the Armed Forces Institute of Pathology, 3rd Edition. McGraw-Hill Book Co., New York.

LYNCH, M.J. *et al.* 1969. Medical Laboratory Technology and Clinical Pathology. W.B. Saunders, Philadelphia.

MCMANUS, J.F.A., and MOWRY, R.W. 1960. Staining Methods, Histologic and Histochemical. Paul B. Hoeber, New York.

PAPANICOLAOU, G.N. 1942. A new procedure for staining vaginal smears. Science *95*, 438–439.

PEARSE, A.G.E. 1960. Histochemistry, Theoretical and Applied, 2nd Edition. Little, Brown and Co., Boston.

PREECE, A. 1965. A Manual for Histologic Technicians, 2nd Edition. Little, Brown and Co., Boston.

SHEEHAN, D.C., and HRAPCHAK, B.B. 1973. Theory and Practice of Histology. C.V. Mosby, St. Louis.